From Climate Finance to Carbon Management - Integrated Pathways for Energy, Resilience, and Planetary Systems

Edited by Katundu Imasiku

Published in London, United Kingdom

From Climate Finance to Carbon Management - Integrated Pathways for Energy, Resilience, and Planetary Systems
http://dx.doi.org/10.5772/intechopen.1008655
Edited by Katundu Imasiku

Contributors
Adanna Henri-Ukoha, Adeyinka Richard Aroyehun, Ahoura Zandiatashbar, Alin Artene, Calum Fitzgerald, Daniel Jemiard Mmasomwayera Sinkula, Elizabeth Ekaette Oladejo, John Allen, Katundu Imasiku, Larisa Ivascu, Lonnie Franks, Marius Pislaru, Serena E. Alexander

First published in London, United Kingdom, 2026 by IntechOpen
IntechOpen is the global imprint of INTECHOPEN LIMITED, registered in England and Wales, registration number: 11086078, 167-169 Great Portland Street, London, W1W 5PF, United Kingdom

For EU product safety concerns: IN TECH d.o.o., Prolaz Marije Krucifikse Kozulić 3, 51000 Rijeka, Croatia, info@intechopen.com or visit our website at intechopen.com.

British Library Cataloguing-in-Publication Data
A catalogue record for this book is available from the British Library

From Climate Finance to Carbon Management - Integrated Pathways for Energy, Resilience, and Planetary Systems
Edited by Katundu Imasiku
p. cm.
Print ISBN 978-1-83769-353-5
Online ISBN 978-1-83769-352-8
eBook (PDF) ISBN 978-1-83769-354-2

Meet the editor

Dr. Katundu Imasiku serves as a Research Fellow at the H. Milton Stewart School of Industrial and Systems Engineering at the Georgia Institute of Technology in the United States. His research focuses on energy systems and climate finance within the Net Zero Freight Program. His work synthesizes financial, engineering, and policy frameworks to promote resilient and scalable clean energy transitions, especially in developing regions. He has taken on leadership roles in renewable energy and green hydrogen initiatives, including the H2Atlas Project, and has a substantial publication record in peer-reviewed journals related to energy systems and sustainability.

Contents

Preface

As climate impacts accelerate and development pressures persist, climate finance has transcended its role as a mere supporting instrument; it is now an essential element in policymaking, building institutional trust, and formulating economic strategies. Funding decisions influence the evolution of energy systems, determine risk coverage, prioritize infrastructure initiatives, and impact community adaptation efforts without interruptions. While investing in clean energy remains crucial for addressing climate change, its progress and stability are increasingly influenced by broader financial factors, including monetary policy, regulatory frameworks, risk assessments, and resource availability to sustain projects amid climate and economic uncertainties.

Concurrently, sectors traditionally not classified as "energy," such as urban planning, transportation, and agriculture, are gaining prominence in shaping climate outcomes. These sectors affect emission trajectories, influence vulnerability to climate-related risks, and determine resilience among households, enterprises, and public systems. Climate finance does not serve as a single tool or funding mechanism; rather, it functions as an interconnected system of decisions and incentives across multiple levels: central banks impacting capital costs, corporations reacting to financial and regulatory signals, development and nonprofit organizations managing climate and disaster risks, governments coordinating infrastructure development and land use, and households adapting within their resources, access, and available support.

Despite growing attention from policymakers and researchers, significant implementation gaps persist. Practitioners often emphasize the need for more robust financing solutions, improved risk-mitigation tools, and more reliable data to guide investment decisions. These issues are becoming increasingly critical as climate variability coincides with inflation, rising debt levels, and increased uncertainty, factors that raise capital costs and weaken the resilience of climate and clean energy initiatives.

This volume, *From Climate Finance to Carbon Management - Integrated Pathways for Energy, Resilience, and Planetary Systems*, examines a crucial domain where climate finance wields considerable influence: the intersection of financial systems, risk mitigation, governance, and practical implementation. It consolidates insights into the operational mechanisms of climate finance within overarching financial frameworks, its role in fostering resilience against risks, and its capacity to catalyze positive transformations across sectors, including infrastructure, planning, and food systems.

The volume further expands this analysis to encompass emerging methodologies in carbon management, including interventions that directly involve Earth system processes. It examines large-scale strategies for regulating the global carbon cycle, particularly within oceanic systems, as supplementary avenues for mitigating atmospheric carbon disparities.

The chapters embody these complementary perspectives. Chapter 2 (Ivașcu, Artene, and Pîslăru) carefully examines the relationship between central banking policies, financial strategies, and sustainability, demonstrating how macro-financial conditions affect business decision-making and the implementation of circular-economy principles. Chapter 3 (Sinkula) underscores the significance of disaster risk financing in Sub-Saharan Africa, highlighting how financial protection mechanisms help sustain clean energy initiatives despite climatic and institutional constraints. Chapter 4 (Alexander and Zandiatashbar) provides insights into multi-level governance in transportation and land-use planning in California, illustrating how cross-institutional collaboration can bolster the effectiveness of climate investments. Chapter 5 (Aroyehun, Henri-Ukoha, and Oladejo) presents climate-smart soybean agriculture in Nigeria, demonstrating how infrastructure, risk financing, and clean energy systems collaboratively enhance resilience and productivity. Lastly, Chapter 6 (Allen, Fitzgerald, and Franks) explores innovative ocean-based approaches to carbon management, acknowledging their promising potential while acknowledging the challenges of scalability, environmental integrity, and governance.

Collectively, the chapters highlight several key insights. Climate outcomes are affected by decisions made within interconnected systems and institutions. Risk remains a critical factor in evaluating the resilience of investments under stress. Governance and coordination are vital to ensuring that financial resources are efficiently translated into actionable implementation. Measurement and data are essential for establishing credibility and fostering adoption. Equity and access continue to present significant challenges, particularly in environments characterized by high climate exposure and limited access to capital, insurance, and supportive infrastructure.

This volume advocates for a cohesive perspective: climate finance, clean energy deployment, and carbon management are interconnected components within broader frameworks dedicated to mitigating climate change. By integrating financial systems, sector-specific strategies, and planetary processes, it facilitates a more comprehensive understanding of how climate solutions can be effectively developed, financed, and sustained in practice.

It is our sincere aspiration that this volume offers both analytical clarity and practical insights to researchers, policymakers, and practitioners seeking to align financial systems with resilient, scalable climate outcomes in an increasingly complex global environment.

Katundu Imasiku
H. Milton Stewart School of Industrial and Systems Engineering,
Georgia Institute of Technology,
Atlanta, USA

Section 1

Introduction

Chapter 1

Introductory Chapter: Climate Finance, Clean Energy and the Evolving Landscape of Climate Solutions

Katundu Imasiku

1. Introduction

The term "climate finance" encompasses various functions that vary by institution and context. In public finance, it manifests in budget choices, managing fiscal risks, and debt patterns, all within the scope of climate exposure [1, 2]. In development finance, this includes concessional capital, blended financial structures, and project preparation mechanisms that aim to reduce investment barriers [3]. In private markets, climate finance is connected to bankability, reliable revenue streams, stable regulatory environments, and risk-adjusted returns, all of which support long-term capital commitments [4, 5]. Disaster risk management includes tools and strategies that ensure quick liquidity after shocks and transfer risks that could threaten public budgets and households [6, 7]. Increasingly, these functions are also being considered alongside emerging carbon management approaches, including interventions that engage directly with Earth system processes.

These differences are significant, reflecting the shift of climate finance toward a framework of rules, incentives, institutions, and risk-sharing mechanisms that shape development priorities, protection strategies, and the capacity to cope with disruptions [8]. As climate impacts become more severe and macro-financial conditions grow more unpredictable, this system is increasingly affecting the deployment of clean energy as well as the resilience of essential infrastructure and livelihoods [7-9]. Furthermore, mobilizing capital, allocating risks, coordinating institutions, and assessing the measurability and legitimacy of outcomes are other key components of climate finance [10, 11].

Increasingly, climate action also involves engagement with Earth system processes, including the global carbon cycle. In this context, approaches such as nature-based solutions, carbon dioxide removal, and large-scale interventions in ocean and terrestrial systems are gaining attention as complementary pathways.

Together, climate finance and clean energy deployment, along with broader system-level considerations, including risk, sectoral implementation, and emerging approaches to carbon management must be examined collectively. By doing so, we can highlight the need for integrated pathways that link financial systems, infrastructure, and planetary processes in addressing climate change. Understanding these integrated pathways requires examining how climate finance has evolved across different institutional contexts and functions.

2. From environmental spending to climate finance systems

Climate finance has progressed through several stages: starting with basic environmental and development expenditures, then adopting market mechanisms for measurable mitigation, and more recently focusing on financial system governance, including risk management, transparency, and capital allocation [4, 5, 10]. Understanding this progression is crucial because current climate finance discussions often mirror the varied institutional backgrounds of participants.

2.1 Climate finance as public spending and development support

The initial climate-related investment is incorporated into broader development goals, such as electrification, water access, agricultural output, and pollution reduction [1, 12]. In many low- and middle-income countries, the main barrier is not a lack of climate ambition but practical constraints, including limited fiscal space, institutional capacity, and necessary infrastructure [2, 9]. When used, climate framing generally serves as an additional justification for investments already prioritized for development and public welfare.

The public spending aspect of climate finance remains crucial. Many adaptation and resilience investments depend heavily on public budgets and service systems, especially in areas with limited household insurance coverage and restricted access to private credit. Consequently, fiscal policy, debt sustainability, and productive allocation are key factors shaping climate outcomes and fundamental constraints on implementation [2].

2.2 Climate finance as a market mechanism and carbon value

This stage focuses on mechanisms that translate emissions reductions into credible value, with increasing emphasis on measurement, reporting, verification, and additionality [1, 13]. While this approach advanced technical disciplines, it also revealed ongoing challenges. Issues with measurement, contested rules, and uneven distribution of benefits can undermine credibility and political support [1, 13]. Moreover, mitigation strategies alone often do not overcome all obstacles to project implementation, especially those related to revenue certainty, system preparedness, and risk sharing [6, 14].

2.3 Climate finance as financial system governance

A third and increasingly important phase positions climate finance within financial system governance. This includes climate risk disclosure, taxonomy development, and the integration of transition and physical risks into investment decisions and regulatory frameworks [10, 15]. This shift expands climate finance beyond project-level funding toward broader questions of market structure, capital allocation, and long-term system transformation. These dynamics influence the cost of capital, access to finance, and regional disparities in investment, ultimately shaping the pace and distribution of clean energy deployment [4, 15]

This reconsideration shifts the focus from the amount of financing to the conditions that make capital investable, sustainable, and scalable, such as access and cost considerations [4, 5]. These factors explain ongoing differences in capital costs across regions and sectors, influencing the speed and distribution of clean energy deployment [5, 14] These conditions become most visible when examining how clean energy projects are financed and delivered in practice.

3. Clean energy as a financing and delivery problem

While technological advances and learning curves are often associated with the shift to clean energy, the larger-scale deployment increasingly depends on financial and institutional factors that affect bankability and feasibility [4, 5]. These constraints are common across regions.

First, the cost of capital is crucial, particularly when country, currency, and regulatory risks raise financing costs [5, 14]. Second, revenue certainty usually influences bankability. Stable revenue streams can come from regulated tariffs, reliable procurement, long-term agreements, or policy backing. When these sources are uncertain, investment often declines, despite the availability of resources and technology [4, 16]. Third, system readiness, such as grid connection, transmission capacity, storage, and permitting, can restrict how effectively capital commitments are turned into operational capacity [14, 17]. Fourth, regulatory durability influences whether long-term assets remain financeable amid economic and political shifts; instability increases risk premiums and reduces investment horizons [4, 5, 10].

These constraints highlight a broader point: clean energy transitions are not just about technology. They also examine whether climate finance can effectively serve as a delivery system that moves projects from initial commitment and planning through construction and operations, while building resilience against shocks [14, 17]. This delivery challenge increasingly extends to carbon management approaches that interact directly with Earth system processes, including large-scale interventions in the carbon cycle, which raise additional demands for governance, measurement, reversibility, and long-term institutional credibility.

In addition, clean energy outcomes are shaped by developments in sectors that influence demand, exposure, and the feasibility of implementation. Transport systems and land-use planning affect energy demand and emissions patterns [8, 11]. Land governance determines sitting, permitting, and vulnerability to climate hazards [12]. Agriculture and food systems, especially in climate-vulnerable regions, rely on enabling infrastructure and risk management tools to support adoption under conditions of uncertainty [2, 9].

4. Volatility as the baseline: Why "finance that delivers" is distinct

The interplay of physical and macro-financial volatility shapes the modern financial environment. Physical risks, including floods, droughts, storms, and heat stress, increase the likelihood of disruptions and asset damage, while restrictive financial conditions raise hurdle rates and constrain public budgets [2, 9, 12]. Under these circumstances, financeability assesses the entire lifecycle of an asset or program rather than only at financial close.

Continuity mechanisms are vital for maintaining delivery during periods of volatility. Instruments like insurance and catastrophe-linked tools help safeguard balance sheets and enable quicker recovery [6, 7]. Contingency financing, pre-arranged liquidity, and structured financial protection can reduce the fiscal and welfare impacts of shocks. Resilient design standards help minimize damage and service disruptions, improving infrastructure reliability during extreme events [9, 12]. The effectiveness of these mechanisms depends on institutional support capacity, procurement, claims management, disbursement systems, and inter-agency coordination [2, 7].

In environments with high exposure and limited budgets, the absence of financial safeguards can undermine progress, transforming long-term plans for clean energy and response measures into ongoing reconstruction efforts [6, 9]. Resilience and disaster-risk finance should be regarded as fundamental components of climate finance systems rather than optional tools.

5. Governance across scales as a multiplier of investment effectiveness

While climate finance can be assessed on a global scale, its implementation is influenced by institutions operating at multiple levels, including national frameworks, subnational planning, sector regulators, utilities, municipalities, companies, and households [2, 4, 9]. This multi-level structure introduces coordination challenges that can either help or hinder outcomes. Usually, national officials establish targets, incentives, and regulatory frameworks, and provide fiscal support or guarantees [1, 2, 4]. Subnational authorities influence land use, transportation systems, and permitting processes, often determining whether investments can proceed on schedule and at a reasonable cost [8, 11]. Utilities and regulators affect grid access, investment strategies, and system reliability, all of which are key to expanding clean energy [3, 14] Firms and financial institutions consider policy credibility, disclosure requirements, and risk pricing when allocating capital [5, 15]. Households and producers choose technologies and practices based on their affordability, access to credit or insurance, and perceived legitimacy [2, 9]

When roles and incentives are aligned, even limited financing can achieve significant results through prompt action and lower project risk. Conversely, conflicting mandates or divided responsibilities may cause delays and increase risk premiums, ultimately weakening well-funded initiatives [3, 11]. Effective governance coordination enhances financial efficiency and goes beyond just administrative issues. These dynamics also frame the broader scope, which includes emerging approaches to carbon management that interact directly with Earth system processes.

6. Measurement, legitimacy, and the politics of "what counts"

Measurement is essential in climate finance as it supports accountability, facilitates learning, allows for comparability, and helps reassure investors [4, 9, 10]. However, this can also lead to difficulties. Complex reporting might surpass local capabilities, standards can be contested between jurisdictions and sectors, and incentives may prioritize easily measurable metrics over those that more accurately reflect delivery and welfare outcomes [1, 4, 10]. Data gaps and varying definitions can undermine trust and increase financing costs, particularly when standardized evidence is necessary [5, 15].

Legitimacy is essential when addressing these technical challenges. Policies and investments that distribute costs and benefits unevenly could provoke political resistance and jeopardize continuity [2]. Similarly, when markets perceive rules as unstable or disclosures as untrustworthy, the cost of capital rises and investment decreases [4, 5, 15]. These factors suggest that climate finance gaps are influenced not just by funding amounts but also by institutional trust, the accuracy of measurement, and equitable design, all of which support sustained long-term progress [12, 15]

7. A delivery-oriented framework for climate finance

A practical interface where climate finance either succeeds or fails, synthesizing diverse sectoral and geographic perspectives, emphasizes the intersection of financial rules, risk management capacity, governance coordination, and concrete delivery [3, 4, 6].

1. **Financial rules and incentives:** These determine eligibility, revenue consistency, and the policy credibility required for bankability [3, 4, 16].

2. **Risk management capacity:** This concerns how physical and financial risks are reduced, priced, shared, and transferred, including the design of financial protection and uncertainty management [3, 4, 16].

3. **Governance coordination:** This includes the alignment of authority, timelines, and implementation responsibilities across institutional levels and sectors [2, 11].

4. **Real-world delivery:** This includes project preparation, construction, operational performance, system integration constraints, and credible monitoring that supports continuity and learning [1, 3, 14].

This framework supports uniform analysis across different sectors by pinpointing the factors that enhance intervention bankability, identifying and managing risks, determining the coordination needed among actors, and establishing methods for monitoring and sustaining outcomes [6]. It also emphasizes that the main reason for delivery failures is often misalignment among these components, rather than just a lack of capital.

8. Concluding perspective

Moving toward effective finance: Climate finance debates tend to emphasize scale, treating capital mobilization as the primary objective. While necessary, scale only creates potential; effective delivery is what transforms that potential into durable mitigation, resilience, and welfare gains [1, 3, 11]. Delivery depends on rules that establish investability, risk frameworks that remain effective during volatility, governance arrangements that coordinate action across levels, and measurement systems that promote legitimacy and facilitate learning [6, 15, 16].

Climate finance now operates through various institutional modalities, including public expenditure, fiscal policy, market-based mechanisms, and financial system governance. It also highlights that achieving clean energy outcomes increasingly relies on how these modalities influence incentives and constraints within a specific context [1, 5, 15]. This work narrows the gap between commitment and construction, between planning and performance, and between finance mobilization and resilient outcomes [6, 11, 14]. With this framing in place, subsequent chapters apply these lenses to examine how alignment across rules, risk allocation, institutions, and measurement can strengthen finance that delivers. This analysis also encompasses emerging approaches to carbon management that interact directly with Earth system processes, including large-scale interventions in the global carbon cycle, which introduce new financing, risk, and governance considerations for their development and deployment.

Acknowledgments

Author acknowledges the use of Grammarly to improve the language in their work.

Author details

Katundu Imasiku
H. Milton Stewart School of Industrial and Systems Engineering, Georgia Institute of Technology, Atlanta, USA

*Address all correspondence to: katunduimasiku@gmail.com

References

[1] UNFCCC. "Report of the Standing Committee on Finance" [Online]. 2022. Available from: https://unfccc.int/documents/618943 [Accessed: 2026-May-25]

[2] IMF. "Fiscal Policies for Paris Climate Strategies-from Principle to Practice" [Online]. 2019. Available from: https://www.imf.org/-/media/files/publications/pp/2019/ppea2019010.pdf [Accessed: 2026-February-04]

[3] Inter-American Development Bank. et al. "2022 Joint Report on multilateral development banks' climate finance." 2023. DOI: 10.18235/0005182

[4] EU Regulation. "Regulation (EU) 2020/852 of the European Parliament and of the Council of 18 June 2020." 2020. Available from: https://publications.iadb.org/en/2022-joint-report-multilateral-development banks climate-finance [Accessed: 2026-February-04]

[5] OECD. Climate Finance Provided and Mobilised by Developed Countries in 2016-2020. In: Climate Finance and the USD 100 Billion Goal. Paris: OECD Publishing; 2022. DOI: 10.1787/286dae5d-en

[6] World Bank. "Disaster Response: A Public Financial Management Review Toolkit" [Online]. 2019. Available from: https://documents1.worldbank.org/curated/en/814381594112608828/pdf/Disaster-Response-A-Public-Financial-Management-Review-Toolkit-Approach-Note.pdf [Accessed: 2026-February-04]

[7] World Bank. "Financial Protection Against Natural Disasters." 2014. Available from: https://documents1.worldbank.org/curated/en/523011468129274796/pdf/949880WP0Box380st0Natural0Disasters.pdf [Accessed: 2026-February-04]

[8] IPCC. "Climate Change 2022 Mitigation of Climate Change." 2022. Available from: https://www.ipcc.ch/report/ar6/wg3/ [Accessed: 2026-February-04]

[9] UNEP. Adaptation Gap Report 2023: Underfinanced. Underprepared. Inadequate investment and planning on climate adaptation leaves world exposed. United Nations Environment Programme. 2023. DOI: 10.59117/20.500.11822/43796

[10] TCFD. "Task Force on Climate-related Financial Disclosures." 2022. Available from: https://www.fsb-tcfd.org/ [Accessed: 2026-February-04]

[11] IPCC. "Climate Change 2023: Synthesis Report." 2023. DOI: 10.59327/IPCC/AR6-9789291691647

[12] IPCC. "Climate Change 2022: Impacts, Adaptation and Vulnerability." 2022. Available from: https://www.ipcc.ch/report/ar6/wg2/downloads/report/IPCC_AR6_WGII_FullReport.pdf [Accessed: 2026-February-04]

[13] ISO. "ISO 14064-1:2018 - GHG Quantification and Reporting Standard" [Online]. Available from: https://www.iso.org/standard/66453.html [Accessed: 2026-February-04]

[14] IEA. "World Energy Outlook 2023" [Online]. 2023. Available from: https://

www.iea.org/reports/world-energy-outlook-2023 [Accessed: 2026-February-04]

[15] Carney MM. "Recommendations of the Task Force on Climate-related Financial Disclosures" [Online]. 2017. Available from: https://assets.bbhub.io/company/sites/60/2020/10/FINAL-2017-TCFD-Report-11052018.pdf [Accessed: 2026-February-04]

[16] NGFS. "NGFS Scenarios for central banks and supervisors." 2023. Available from: https://www.ngfs.net/en/publications-and-statistics/publications/ngfs-climate-scenarios-central-banks-and-supervisors-phase-iv [Accessed: 2026-February-04]

[17] IEA. "Net Zero Roadmap: A Global Pathway to Keep the 1.5°C Goal in Reach - 2023 Update" [Online]. 2023. Available from: https://iea.blob.core.windows.net/assets/8ad619b9-17aa-473d-8a2f-4b90846f5c19/NetZeroRoadmap_AGlobalPathwaytoKeepthe1.5CGoalinReach-2023Update.pdf [Accessed: 2026-February-04]

Section 2

Foundations - Climate Finance and System Transitions

Chapter 2

Green Finance Strategies: Integrating Circular Economy Principles and Central Bank Policies for Sustainable Development

Larisa Ivascu, Alin Artene and Marius Pislaru

Abstract

The principles of the circular economy are being adopted by more and more organizations. The transition to the circular economy involves regulations, strategies, and policies. Banking systems highlight concerns about sustainability in the context of the consolidation of financial strategies. This chapter examines the interplay between monetary policies set by central banks, business financial strategies, and the pursuit of sustainability, with a particular focus on integrating circular economic principles. It is directed in two directions using qualitative and quantitative evaluation. Qualitative evaluation is used to assess specialized literature. Quantitative evaluation analyzes the impact of monetary policies set by central banks on enterprise strategies and sustainability and is based on a structured survey applied to a representative sample. The hypotheses were tested on a sample of 530 responses. Qualitative evaluation is performed on an inventory of specialized literature. It begins by analyzing recent decisions of central banks and the monetary policy responses to post-pandemic inflation. Our chapter also explores opportunities for companies to align their financial strategies with sustainability goals, emphasizing the importance of integrating circular economic principles. Furthermore, it briefly touches on the challenges and opportunities in implementing sustainability initiatives, emphasizing the need for collaboration across sectors to drive sustainable finance and address climate-related risks effectively. The results of the quantitative research show that there is a positive correlation between central policies and companies. The level of information influences the knowledge. The chapter ends with conclusions and future research directions.

Keywords: monetary policy, financial policies, circular economy model, cost of money, sustainability, management, post-pandemic, EU Central Bank, Basel III

1. Introduction

The growing importance of sustainability positions central bank policies as catalysts for change and underscores the transformative potential of green finance strategies to navigate the complexities of the post-pandemic era. Approaching the 17 Sustainable Development Goals (SDGs) is becoming a concern for more organizations. Following the unprecedented challenges posed by the pandemic, the intersection of monetary policy, sustainability, and economic resilience has become a growing concern for governments and financial organizations around the world [1]. As Europe's economies grapple with the disruptive impact of the pandemic, the resurgence of pandemic inflation provided central banks with a stress test of their ability to maintain price stability and protect long-term economic sustainability.

In the specialized literature, we have witnessed a remarkable shift in the perception of sustainability, and the paradigm has evolved from a niche concern to a mainstream imperative across sectors. The increased awareness from the G20 summits and global events on climate change, resource depletion, and social inequities has underscored the urgency of transforming our current linear economic models to protect the planet and ensure intergenerational equity. In our pursuit of sustainable development, the intersection of central bank policies and green finance strategies emerges as a cornerstone of economic resilience and environmental stewardship. As European economies recover from the pandemic, the need for green finance integration in ECB policies grows stronger. Central banks have a vital role in this process. They embed sustainability into monetary policy and support green investment initiatives. They also conduct climate stress tests to assess financial risks. In addition, they promote sustainable finance standards and encourage cooperation and knowledge sharing among institutions (see **Figure 1)** [2]. **Figure 1** illustrates the interconnectivity between central bank policies, sustainable investments, and circular economy principles, as conceptualized by the authors. The visual model positions the central bank at the core of a multidimensional framework that bridges monetary policy instruments with sustainability-driven economic mechanisms.

In the upper section, the figure presents the key monetary tools – such as interest rates, asset purchases, and cost-of-money frameworks – that define the operational scope of central bank policies. These instruments directly shape market liquidity, borrowing costs, and investment incentives.

The middle and lower sections link these policy mechanisms to sustainability and circular economy domains, illustrating how central banking decisions can encourage sustainable investments, renewable energy development, and green infrastructure financing. Similarly, policies aligned with circular economy principles – including waste reduction, resource efficiency, and lifecycle sustainability – contribute to broader financial and environmental resilience.

Overall, the figure emphasizes the systemic role of central banks in integrating economic stability with environmental sustainability, promoting a policy ecosystem that supports both financial soundness and long-term circular growth.

At the heart of the sustainable development European Union (EU) Parliament agenda should lie the concept of green finance strategies, which set out to mobilize financial resources toward environmentally sustainable and socially inclusive European investments [3]. By integrating the principles of a circular economy, resource efficiency, and climate resilience into financial decisions, green finance strategies offer a canvas and a pathway toward a more sustainable and resilient

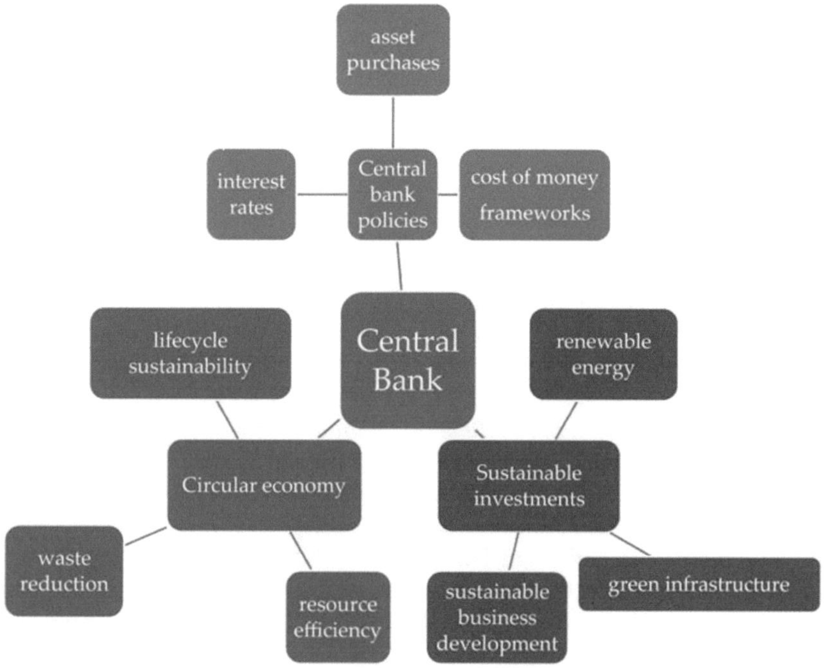

Figure 1.
Interconnectivity of central bank policies, sustainable investments, and circular economy principles (authors' approach).

economic capital market. Central banks play a pivotal role in catalyzing the transition toward a greener and more sustainable financial system. They have been proactive in integrating sustainability considerations into their monetary policy frameworks and have attempted to navigate rising inflation through various strategies tailored to their respective economic contexts and policy frameworks.

Although some economies have tightened monetary policy in response to inflationary pressures and volatile, fragile expectations, central banks in advanced economies have been cautious in adjusting policies, reflecting a delicate balance between price stability and the objectives of economic recovery by incorporating environmental factors into risk assessments, asset purchase programs, and collateral frameworks.

The growing importance of sustainability positions central bank policies as drivers of change, emphasizing the transformative potential of green finance strategies in navigating the post-pandemic complexities. Amid unprecedented challenges, the intersection of monetary policy, sustainability, and economic resilience should become a European focal point [4].

The resurgence of pandemic inflation stressed the commitment of central banks to maintain price stability during economic disruption. Developing from a marginal concern, sustainability has gained mainstream academic and economic recognition, driven by increased awareness of climate change and social inequities [5]. This shift underscores the urgency of transitioning from linear economic models to protect the planet and ensure intergenerational equity through circular economic models. The integration of green finance principles into European Central Bank

(ECB) policies gains prominence in post-pandemic recovery, aligning with central banks' roles in advancing sustainability through monetary policy, supporting green finance initiatives, and fostering collaboration.

At the core of the EU's sustainable development agenda, green finance strategies and mobilizing resources toward sustainable investments should be the focus. These strategies, incorporating circular economy principles, will offer pathways to a more resilient economic capital market.

Central banks in Europe should catalyze the transition to greener financial systems, proactively integrating sustainability [6] into monetary policy frameworks. In addressing inflationary pressures, they should strike a balance between price stability and economic recovery objectives. This cautious approach can reflect contextual considerations and the evolving landscape of sustainable finance in our near future.

The innovative part of this article is underlined by the comprehensive review of specialized literature and the development of a survey that addressed experts in monetary policies at the international level. The respondents were people with leadership positions in the area of monetary policies. This research advances the specialized literature by defining a framework for the types of decisions that need to be made in the field of monetary policy and sustainability. It examines the interplay between monetary policies set by central banks, financial strategies of businesses, and the pursuit of sustainability, with a particular focus on integrating circular economic principles. This research also explores opportunities for companies to align their financial strategies with sustainability goals, emphasizing the importance of integrating circular economic principles. Furthermore, it briefly touches upon the challenges and opportunities in implementing sustainability initiatives, emphasizing the need for collaboration across sectors to drive sustainable finance and address climate-related risks effectively. The quantitative research results show that there is a positive correlation between central policies and enterprises.

The comprehensive analysis and innovative nature are supported by the design of the research concept, which includes addressing the impact of EU policies, circular economic principles, and sustainability on sustainable investments. At the Romanian level, there is no such approach, and the results help entrepreneurs make improved management decisions to increase organizational competitiveness.

This chapter examines the interplay between monetary policies set by central banks, business financial strategies, and the pursuit of sustainability, with a particular focus on integrating circular economic principles. The results obtained contribute as important factors in the decision-making process and address this gap that is not adequately covered in the specialized literature.

2. Evaluation of the theoretical implications of the concepts addressed

2.1 The role of central banks in influencing the availability and cost of money for sustainable investments

Central banks in Europe shape monetary policy by adjusting interest rates to influence borrowing, conducting open market activities to regulate the money supply, creating reserve requirements according to Basel III [7], providing forward guidance to shape expectations and behavior, implementing asset purchase programs to stimulate investment, and acting as lenders of last resort to stabilize financial markets during crises.

These policies promote green finance initiatives by adjusting interest rates to incentivize green investments, conducting open market operations to lower borrowing costs for green projects, providing forward guidance to signal support for sustainability, including green assets in asset purchase programs to stimulate investment in environmentally friendly initiatives, and acting as lenders of last resort to ensure market stability and continued investment in green projects during crises [8]. Basel III focuses mainly on strengthening the resilience of banks and the banking sector.

Although Basel III did not directly address sustainable investments, its principles and regulations have the potential to influence banks' engagement with sustainable finance. The finalized Basel III reforms, through its five-year full implementation timeframe, introduce several improvements and refinements to the regulatory framework established by Basel III, particularly in the standardized approach to credit risk and market risk [8]. The Basel Committee sets standards, while central banks and supervisors implement them.

Sustainable investments can be boosted by supporting SMEs engaged in green initiatives or investment-grade corporates with strong environmental performance, thereby promoting sustainability in the corporate sector, while lower risk weights for mortgages and commercial real estate projects with sustainable attributes may encourage banks to finance environmentally friendly buildings and developments.

Finalized Basel III reforms replace flat risk weights for subordinated debts and equity with risk-sensitive weights that reflect the underlying risk profiles of these exposures by aligning the risk weights with the riskiness of subordinated debts and equity investments, and encourage banks to finance green initiatives and environmentally responsible businesses. Lower risk weights for equity holdings made pursuant to nationally legislated programs or investments in sustainable projects may incentivize banks to support sustainability objectives [8]. The Basel framework provides a more consistent identification of material risk factors across banks and helps them assess and manage risks associated with investments in green initiatives and sustainable projects, enhancing their ability to support sustainability objectives while maintaining prudent risk management practices (see **Figure 2**).

Central banks can monitor improvements in credit risk assessment, market risk management, and regulatory frameworks, and contribute through monetary policy to boost incentives, develop tools to allocate capital prudently, and support sustainable finance initiatives.

By aligning risk weights, capital requirements, and risk management practices with sustainability objectives, future monetary policies may contribute to the promotion of green initiatives and sustainability within the banking sector [9]. Basel IV can influence the role of central banks in ensuring financial stability by strengthening capital adequacy, enhancing risk management practices, promoting market stability, and emphasizing macroprudential regulation.

Beyond regulatory frameworks, central banks wield significant influence over the availability and cost of money, playing a crucial role in shaping the financial landscape and directing capital flows toward sustainable investments. Through monetary policy tools such as interest rate adjustments, open market operations, and reserve requirements, central banks can incentivize or disincentivize lending for

Figure 2.
Enhancements in the regulatory framework for banks, supporting green initiatives and sustainability (authors' approach).

specific purposes, including investments in environmentally friendly projects and sustainable initiatives.

TheECB's response to inflationary pressures should be intertwined with its broader objectives of promoting sustainable growth and green finance initiatives. It is essential to maintain price stability while supporting economic resilience to geopolitical uncertainties by adjusting its policy stance to address inflationary pressures, while providing clarity and stability to financial markets, which will be essential for promoting investments in environmentally sustainable projects [10].

Central banks' asset purchase programs will play a crucial role in supporting the transition to a circular economy by providing liquidity to financial markets and facilitating access to credit for companies engaged in sustainable practices, ensuring future green investments flourish and contribute to a more sustainable economic model.

This year, the ECB committed to ensuring financial stability while adapting to evolving economic conditions and promoting sustainable development, including the transition to a green economy. The ECB's decision to adjust the deposit facility rate (DFR) and provide liquidity through a broad mix of instruments reflects its efforts to steer short-term money market rates in line with monetary policy decisions while supporting the transition to a more sustainable financial system. **Figure 3** represents a circular model of the monetary policy framework, structured into four interdependent components: efficiency, effectiveness, robustness, and flexibility. These components outline the following details of the interdependence:

- Effectiveness – refers to the alignment of monetary policy with the provisions of the European Union treaties and to maintaining a close correlation between interest rates and economic policy.

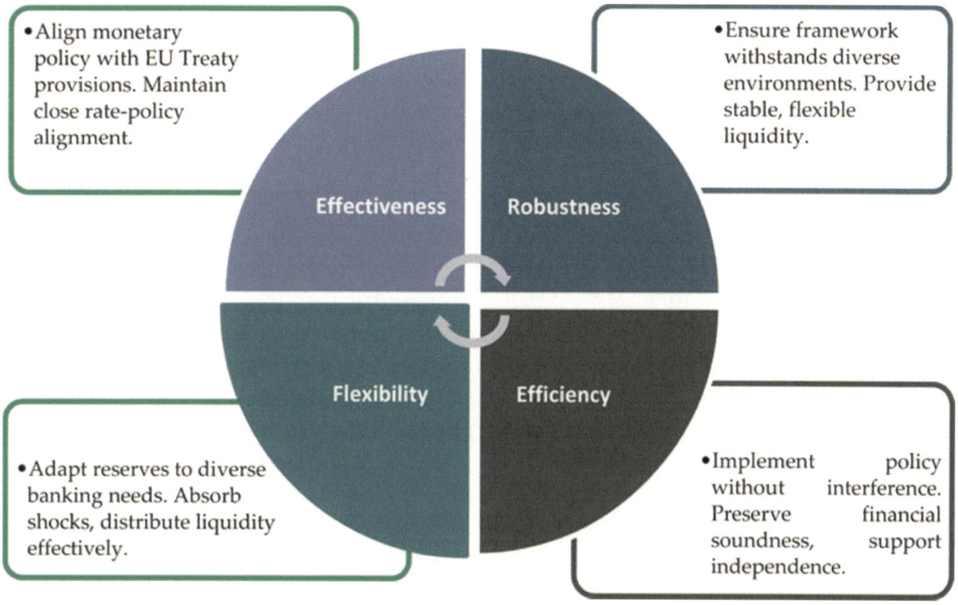

Figure 3.
Future principles for monetary policy assumed by the Governing Council of the European Central Bank [11].

- Robustness – emphasizes the need for a framework capable of withstanding diverse economic conditions, ensuring stable and flexible liquidity.
- Flexibility – aims to adapt bank reserves to the varied needs of the financial system, enhance the capacity to absorb economic shocks, and distribute liquidity efficiently.
- Efficiency involves implementing monetary policy without external interference, maintaining financial soundness, and protecting institutional independence.

The circular arrows suggest the interdependence and dynamic balance between these four dimensions, indicating that an effective monetary policy must combine effectiveness, robustness, flexibility, and operational efficiency in a balanced manner.

By reducing the spread between the rate on main refinancing operations (MROs) and the DFR, the ECB aims to incentivize market activity and provide incentives for banks to seek market-based funding solutions, thus promoting investments in green initiatives and sustainable development projects. Moreover, the ECB's principles for guiding the implementation of monetary policy emphasize effectiveness, robustness, flexibility, efficiency, and openness to support the smooth functioning of markets and contribute to the transition to a green economy. By incorporating climate change-related considerations into its operational framework, the ECB recognizes the importance of addressing environmental challenges while pursuing its main objective of price stability.

Following the inventory of the specialized literature, a first hypothesis can be structured.

H1: Central monetary policies influence the strategies of enterprises and sustainable development.

2.2 The principles of circular economy and the financing of sustainable business development by promoting investments in renewable energy

The circular economy and sustainability are concepts addressed at the macroeconomic level. The circular economy involves a production and consumption model based on a series of functions that contribute to the maintenance of products or materials for as long as possible. The transition from a linear economy to a circular economy implies the efficiency of activity, the reduction of the use of natural resources, and the decrease of the impact on the environment [12].

The functions of the circular economy are presented in the following figure. The evolution of these functions occurs in accordance with economic dynamics and new technological trends. The application of these functions contributes to the transition from a linear economy to a circular economy (see **Figure 4**).

Refuse	• the use of environmentally friendly raw materials that do not affect the environment.
Rethink	• improving the production process and the approach, by thinking in terms of the circular economy.
Reduce	• reducing the amount of resources used in production because less is more efficient.
Reimagine	• improving the image to increase the impact of the product.
Reuse	• reuse through resale, rental or donation.
Repair	• activities that contribute to contribute to bringing the product to the basic functional use.
Refurbish	• activities that contribute to restoring an old product to a new one. remaking an old product and turning it into a new one.
Remanufacturing	• extensive activities that contribute to the improvement of the product station.
Redesign	• it involves a process of rethinking the flow and an adequate design.
Reinvention	• assigning a new meaning, the total change of technologies and the innovative materialization of products.
Recycling	• the use of materials in other processes.
Recovery	• the recovery of parts of the products.
Reduction	• decrease in the amount of materials used in organizational and production processes.
Repurpose	• the use of parts of the product for the manufacture of a totally new one.

Figure 4.
The 14 Rs of circular economy [13, 14].

Reuse involves the elimination of materials that are not environmentally friendly and could pollute considerably. Rethinking processes involves strategic decisions regarding the improvement of production approaches in terms of the circular economy. Reducing the amount of raw materials is important in this theoretical framework. The concept of reimagine can be defined as the strategic process by which an organization rethinks and reconfigures its business models, operational processes, and value proposition to be aligned with the principles of circularity and sustainability. Reimagine involves a paradigm shift – not just linear optimization of resources, but reconceptualizing the entire value chain so that materials, energy, and products are used for as long as possible, minimizing waste and reintegrating resources into productive cycles. Reuse refers to extending the life of products, components, or materials by using them repeatedly in the same form and for the same function for which they were originally designed, without requiring a complex industrial transformation process. Reuse involves maintaining the original value and quality for as long as possible, thus reducing the need to extract new resources and the amount of waste generated. In the context of the circular economy, repair is the process by which a defective or damaged product is returned to its original operating condition by replacing or rebuilding defective components, thus extending its useful life. Repair focuses on maintaining the functional value of the product with minimal additional resource consumption, avoiding complete replacement and reducing waste. Refurbishment refers to the process of returning a used product to a condition close to its original, both functionally and esthetically, through a combination of cleaning, repair, component replacement, and possibly technological updating. Unlike repair, which strictly aims to remedy a specific defect, refurbishment involves a broader intervention, with the aim of improving the performance, appearance, and lifespan of the product. Remanufacturing is the process by which a used or end-of-life product is completely rebuilt, often to the same or higher standards as the original, using both recovered and new parts. Unlike refurbishment, which focuses on general reconditioning, remanufacturing involves the complete disassembly of the product, the inspection of each part, the replacement or remanufacture of defective components, and the reassembly of the product according to the manufacturer's original (or even improved) specifications. Redesign is the process of designing or re-imagining products, services, or processes so that they are compatible with the principles of circularity throughout their entire life cycle. The main goal is to prevent waste and maximize the value of resources by integrating solutions that facilitate reuse, repair, upgrade, remanufacturing, or recycling. Reinvention refers to the process of fundamentally transforming a product, service, or business model so that it is fully realigned with the principles of circularity, not just partially optimized. Unlike redesign, which often adjusts the structure or functionality of an existing product, reinvention involves rethinking from scratch the purpose, mode of production, use, and value recovery, often leading to radically different solutions from traditional ones. Recycling refers to the process by which materials from products that have reached the end of their useful life are collected, processed, and transformed into secondary raw materials that can be used in the manufacture of new products. The recovery of resources or energy from waste that cannot be reused or recycled effectively includes processes such as incineration with energy recovery or other processes for partially obtaining the components of the initial product. Reduction involves reducing the volume of resources, materials, or energy consumed in production and use, as well as the amount of waste generated. Reuse for another purpose, or repurposing, refers to the use of a used product or material for a function different from the original one, without transforming it through major industrial processes.

The strategies that should target these functions can be divided into three categories [12, 15, 16]:

- Smarter use and innovative approaches: refuse, rethink, reduce.
- Life cycle improvement: reimagine, reuse, repair, reduce, repurpose, refurbish, remanufacture, redesign, reinvention.
- Efficient use of materials: recycling, recovery.

The most important principles can be systematized:

- Reducing the amount of greenhouse gases and pollution.
- The circulation of products and the efficiency of the number of raw materials used.
- Improving the quality of nature and the optimization of economic, social, and technological values.
- Minimizing the extraction of natural resources.

Following the inventory of specialized literature, a second hypothesis can be structured.

H2: Investment policies in renewable energy should be based on the principles of the circular economy and investments for sustainable development.

2.3 Evaluation of strategies and initiatives implemented to support the promotion of sustainable investments in renewable energy

The principles of the circular economy are addressed by the G20 countries (the 20-member group is the intergovernmental forum that includes 19 sovereign countries, the European Union, and the African Union) and the Conference of the Parties (COP) conference [17]. The G20 countries have an opportunity to propose ways to develop and implement strategies and initiatives that support the promotion of sustainable investments in renewable energy by proposing instruments that promote green energy. The international climate conference is organized annually, pursuing the established objective of ensuring global net zero by mid-century and keeping a maximum of 1.5°C degrees of warming within reach [16]. The promotion of green investments [15, 18–20] should address a series of imperatives that support these global concerns:

(1) Strategies: The development and clear communication of strategies that include concerns about renewable energy. The investment offer should expand internationally and align with the concerns of each entity.

(2) Policies and practices: Investors should feel confident that all investments they make will be recovered and contribute to adding value. At the same time, circular economy and sustainability practices should be promoted. This category can include the replenishment of energy markets, risk management and investment recovery, financial policy support, access to financing programs, access to innovative technologies, and the impact of development.

(3) Promotion: The development of promotional frameworks that contribute to the establishment of links between actors and the consolidation of collaborative relationships.

In 2017 and 2018, there were decreases in net capital flows in renewable energy projects of 1% and 3%, respectively, and a decrease in their efficiency was recorded. Investments may decrease, and in this way, there could be deficiencies in achieving the SDGs and the Paris Agreement for climate change. Investments in renewable energies do not only refer to sustainability and the circular economy but also to financing, because the costs of implementing complex projects are high. The future of renewable energies concerns the innovative technologies that will appear and the actors involved in implementation and financing approaches.

At the level of 2021, renewable energy represented 21.90% of the energy consumed, and in 2022, it increased very little, reaching a percentage of 22%. Greenhouse gases decreased by over 30% between 1990 and 2021.

Climate neutrality is an objective of the European Green Deal [19]. This approach aims to:

- Zero greenhouse gas emissions by 2050.
- Economic growth without increasing the use of resources.
- Without negative effects on people and workplaces.

This transition involves new jobs, new business models, new target markets, and new technologies. Investments in green energy generate a series of benefits, both for the organization and for society. The benefits for the organization should be expressed in terms of profit so that these activities are attractive to them. Among the benefits of the transition, there are several, including reducing energy costs, the development of new jobs, the improvement of the level of technology, the enhancement of safety levels, the improvement of quality of life, the development of local communities, and economic growth (see **Figure 5**). Green investments generate a series of interconnected benefits that simultaneously contribute to sustainability and increased organizational competitiveness. Reducing energy costs by adopting efficient technologies and renewable sources not only optimizes economic performance but also strengthens the resilience of organizations in the face of volatility in energy markets. Creating new jobs stimulates the development of human capital and the diversification of skills, supporting the transition to a low-carbon economy. Improving the technological level facilitates innovation and increases productivity, while enhancing safety

Figure 5.
The benefits of investments in green energies (authors' approach).

levels reduces operational risks and protects the health of employees. On a social level, green investments contribute to improving the quality of life by reducing pollution and enhancing the environment, as well as fostering the development of local communities through infrastructure, economic opportunities, and civic engagement. All these effects converge toward a positive macroeconomic impact, stimulating sustainable economic development and providing organizations with a sustainable competitive advantage based on innovation, social responsibility, and strategic adaptability.

The share of energy from renewable sources, at the level of the European Union, shows an increasing evolution during the period 2013–2022. In recent years, it has registered higher growth, reaching a percentage of 23.047% in 2022 (see **Figure 6**; Eurostat, 2022).

Figure 7 shows the situations of some selected countries that, for the year 2022, registered more than 40% energy from renewable sources. Among these countries are Sweden, Finland, Latvia, Denmark, and Albania.

Investments in renewable energy also involve a series of risks and barriers. Among the most significant barriers are the lack of information, high costs of implementation, inadequate infrastructure, lack of technologies, and shortage of qualified personnel. Among the risks that can be identified, technological risks, regulatory uncertainty, market dynamics, and investment risks are particularly notable [13, 14, 17, 21–24].

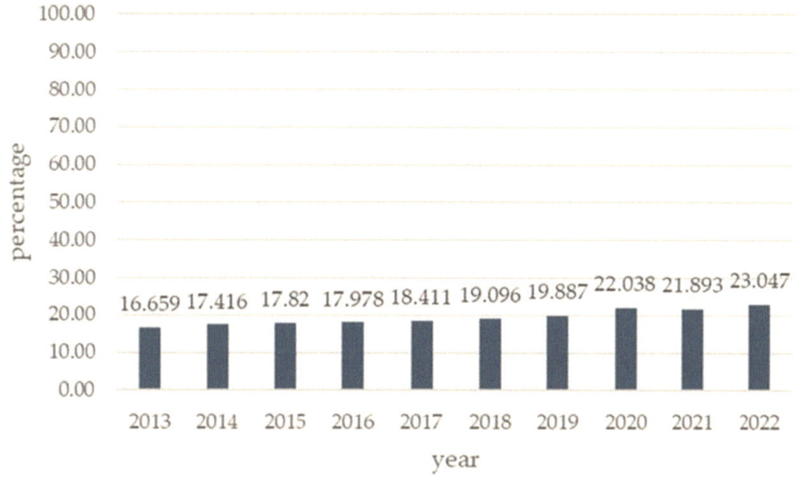

Figure 6.
Share of energy from renewable sources (percentage) for the European Union – 27 countries (from 2020; Eurostat, 2023).

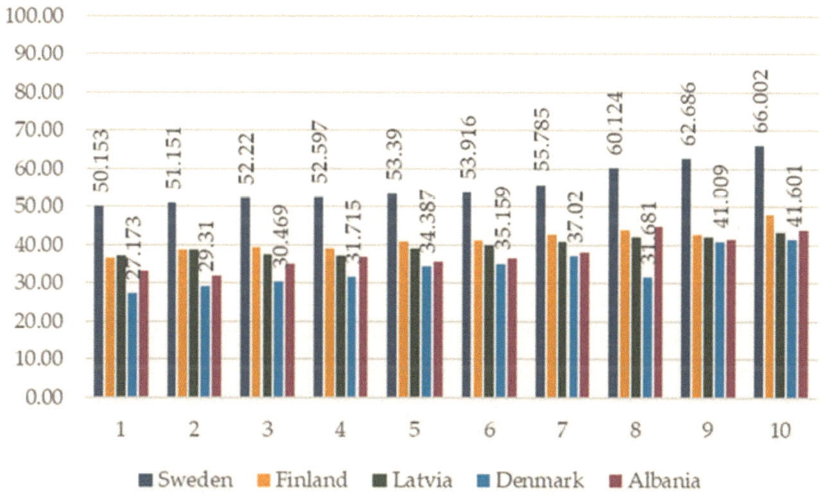

Selected countries	2013	2014	2015	2016	2017	2018	2019	2020	2021	2022
Sweden	50.153	51.151	52.22	52.597	53.39	53.916	55.785	60.124	62.686	66.002
Finland	36.63	38.633	39.23	38.943	40.857	41.185	42.807	43.939	42.854	47.886
Latvia	37.037	38.629	37.538	37.138	39.008	40.019	40.929	42.132	42.098	43.316
Denmark	27.173	29.31	30.469	31.715	34.387	35.159	37.02	31.681	41.009	41.601
Albania	33.167	31.856	34.913	36.953	35.776	36.572	38.042	45.015	41.389	44.076

Figure 7.
Share of energy from renewable sources (percentage) for selected countries that registered a considerable increase (Eurostat, 2023).

3. Research methodology

This research uses qualitative and quantitative research methods. The qualitative method is corroborated by the empirical experience of the authors and presents a review of the specialized literature with an emphasis on the evaluation of data series. The quantitative evaluation uses a questionnaire as an investigation tool. The questionnaire was administered online, and a validated sample of 530 people was obtained. The responses obtained were quantitatively modeled using statistical analyses. The questionnaire was designed based on the research directions of the circular economy, monetary policies, and organizational sustainability. It was structured into three categories of items, with each category including 10 items built on the principles of the above-mentioned concepts. The questionnaire was administered online, and the responses of the respondents were statistically analyzed using SPSS. Variables were selected following an in-depth interview with experts in the field. The sampling frame was composed of monetary policy experts, mostly in leadership positions in financial and regulatory institutions. The questionnaire was distributed online using a random technique. The structure of the questionnaire is:

- Category A – Monetary policies
  Items on knowledge and perception of monetary policy instruments (interest rates, minimum reserves, asset purchases, sustainability integration, etc.).

- Category B – Organizational strategies and sustainability.
  Items on the existence of ESG objectives, circular economy integration, sustainability reporting, green investment initiatives, and more.

- Category C – Renewable energy and circular economy investment policies.

- Items on familiarity and application of the "14 R" principles, renewable investments, climate risks, and barriers to green finance.

The reliability of the items was tested by Cronbach's α coefficient for each of the three measured constructs (Monetary Policies, Organizational Strategies and Sustainability, Renewable & Circular Investments). All α values were above the threshold of 0.70, indicating good internal consistency of the scale (Monetary Policies: $\alpha = 0.82$, Organizational Strategies and Sustainability: $\alpha = 0.85$, Renewable & Circular Investments: $\alpha = 0.87$, and Total Questionnaire: $\alpha = 0.89$). Factor validity was verified by an exploratory factor analysis (EFA). The KMO test had a value greater than 0.80, and the Bartlett test was significant ($p < 0.001$), indicating the adequacy of the data for factor analysis. EFA analysis with Varimax rotation extracted three main factors, corresponding to the theorized constructs, which together explained approximately 65% of the total variance. Then, a confirmatory factor analysis (CFA) was performed using the maximum likelihood estimation method. The three-factor model presented good fit indices: $\chi^2/df < 3$, Comparative Fit Index (CFI) and Tucker-Lewis Index (TLI) > 0.90, and root mean square error of approximation (RMSEA) < 0.08. These results support the convergent and discriminant validity of the measured constructs. It can be concluded that the questionnaire has solid psychometric properties.

4. Results and discussion

The first hypothesis was tested using the Spearman correlation test, measuring the strength and direction of the relationship between the variables.

Correlations				
			Investment policies in renewable energy	**Principles of the circular economy and investments for sustainable development**
Spearman's rho	Investment policies in renewable energy	Correlation coefficient	1.000	0.876
		Significance (two-tailed)	.	0.001
		N	530	530
	Principles of the circular economy and investments for sustainable development	Correlation coefficient		
		Significance (two-tailed)	0.001	.
		N	530	530

Table 1.
Spearman's rank correlation between investment policies in renewable energy, principles of the circular economy, and investments for sustainable development.

Organizational strategies and sustainability were used as independent ordinal variables, and monetary policies were used as dependent ordinal variables, measured on a scale from 1 representing "not at all familiar" to 7 representing "very familiar" (see **Table 1**). Spearman's ρ is appropriate. The choice of this method is justified by the fact that the variables are measured on ordinal Likert scales and may have non-normal distributions. Unlike Pearson's r, Spearman's ρ is based on ranks and evaluates monotonic association between variables without assuming linearity or normality. In addition, the coefficient does not require the designation of a "dependent" and an "independent" variable, treating both variables symmetrically. To test the robustness of the results, we estimated an ordered logit model (alternatively, OLS on composite indices) in which we included, as control variables, the role of the respondents, the level of education, the country, and the size of the firm. The results confirm that monetary policies influence organizational strategies, and the principles of the circular economy are positively associated with green investment policies, independent of the characteristics of the respondents.

Following the statistical analysis using the Spearman correlation coefficient, it can be observed that there is a strong and statistically significant positive relationship between "monetary policies," "organizational strategies," and "sustainability." This result emphasizes that the more solid and stable the central monetary policies are, the more the strategies of the enterprises and their sustainable development position them as competitive organizations. A p-value < 0.001 was obtained. This means that the correlation is very significant, reducing the probability that this association is the result of random variation in the data (see **Table 2**).

In the same way, we tested the second hypothesis using the Spearman correlation test. The results are presented in **Table 2** and validate the correlation between the two variables.

The 14 Rs are important for addressing the principles of the circular economy, and quantitative statistical analysis shows a strong link between investment policies

Correlations				
			Organizational strategies and sustainability	**Monetary policies**
Spearman's rho	Organizational strategies and sustainability	Correlation coefficient	1.000	0.789
		Significance (two-tailed)	.	0.001
		N	530	530
	Monetary policies	Correlation coefficient		
		Significance (two-tailed)	0.001	.
		N	530	530

Table 2.
Spearman's rank correlation between monetary policies, organizational strategies, and sustainability.

in renewable energy, principles of the circular economy, and investments for sustainable development. This means that organizations need to address these 14 Rs to increase their involvement in sustainable development.

Following the statistical analysis using the Spearman correlation coefficient for this second hypothesis, a strong correlation was identified. Therefore, the more solid and competitive investments in green energy, the more investments in the circular economy and sustainability are consolidated and aligned with the global approach. During the survey, it was found that people who did not have management positions were still retained in green investments. The level of education and training of the respondents was an important aspect in the analysis carried out.

To consolidate the practical survey, we tested "Management level" and "Competitiveness at work" using the Chi-square test. This test was applied because it is a statistical test that analyzes the relationship between two categorical variables. If there is a significant relationship, then we can say that there is a significant association between the variables (see **Table 3**).

The analysis of the specialized literature highlights the importance of green investments and the benefits obtained through well-implemented strategies, while the quantitative analysis presents Spearman's rank correlation between monetary

Chi-square test			
	Value	df	Asymptotic significance (two-sided)
Pearson chi-square	419.789	5	0.000
Likelihood ratio	567.890	5	0.000
Linear-by-linear association	3.675	1	0.053
N of valid cases	530		

Table 3.
Chi-square test results for the association between "management level" and "competitiveness at work".

policies, organizational strategies, and sustainability. The strong relationship between the two variables reinforces the importance of organizations developing green strategies and improving their activities to reduce greenhouse gas emissions and promote practices that are beneficial for society.

The results obtained from the application of the Chi-square test are presented in **Table 3**. These results show a significant association between the "Management level" and the "Competitiveness at work," with a Pearson Chi-square value of 419.789, df = 5, and a significance level of $p < 0.001$. This fact tells us that the differences observed between occupational groups are not random. At the same time, it is confirmed that the management level contributes to increased competitiveness at work and to the development of sustainable decisions. A χ^2 analysis showed a significant association between management level and familiarity with monetary policies (Pearson $\chi^2(4) = 18.42$, $p < 0.01$; Likelihood Ratio $\chi^2(4) = 17.95$, $p < 0.01$). Expected frequencies were ≥ 5 in all cells. Analysis of standardized residuals showed that senior managers reported significantly higher levels of familiarity than expected (residual = + 2.6), while respondents at lower levels reported lower levels (residual = –2.1). Post hoc tests with Bonferroni correction confirmed the differences between management levels.

Finally, the level of competitiveness is closely linked to the level of organizational management. The approach of future principles for monetary policy and the implementation of green principles contribute to improved management by increasing the level of knowledge and alignment with the business environment.

Therefore, the two hypotheses have been validated, and it can be stated that monetary policies influence organizational strategies and sustainability. At the same time, circular economy policies and sustainability investments are directly proportional to the central investments proposed for green energy.

The ECB initiatives on the green transition and the integration of environmental, social, and governance criteria into monetary and financial supervision policies have the potential to significantly influence the strategic behavior of organizations. Through mechanisms such as targeting asset purchase programs toward green bonds, including climate risks in financial stability assessments, and incentivizing commercial banks to finance sustainable projects, the ECB can reduce the cost of capital for green investments and create a competitive framework favorable to firms with high sustainability performance. These financial incentives and public policy signals can accelerate the adoption of clean technologies, improve energy efficiency, and the integration of circular economy principles, which, in the medium and long term, can translate into competitive advantages: strengthened reputation, attraction of responsible investors, and preferential access to markets sensitive to circular economy criteria. Thus, the ECB's initiatives not only support European climate goals but can also act as a catalyst for the strategic transformation of organizations, combining sustainability with increased economic performance.

5. Conclusions

The circular economy is a solution for the efficient use of resources. In the important concerns of sustainable development, the EU Parliament agenda should focus on the concept of green finance strategies, which aim to mobilize financial resources toward environmentally sustainable and socially inclusive European investments. Central banks in Europe shape monetary policy by adjusting interest rates to influence borrowing, conducting open market operations to control the money supply, and creating reserve requirements [17, 23]. The results obtained show that there is a strong correlation between central policies and organizational strategies. The quality of decision-making is influenced by the competitiveness of the workplace.

Central banks are tasked with balancing, maintaining price stability, fostering economic growth, and supporting sustainability initiatives. Through a multifaceted approach, central banks can influence the availability and cost of money for sustainable investments, using tools such as monetary policy adjustments, forward guidance, and asset purchase programs. The recent adjustments to the ECB's operational framework underscore the ongoing evolution in central bank strategies to adapt to changing economic conditions and achieve policy objectives effectively. Central banks should refine their approaches and integrate sustainability into monetary policy frameworks to reflect the growing recognition of the interconnectedness between financial stability, economic resilience, and environmental sustainability.

Monetary policies influence organizational strategies and sustainability. At the same time, circular economic policies and investments in sustainability are directly proportional to the central investments proposed for green energy. To strengthen organizational competitiveness, the principles of sustainability and the green economy should be considered and implemented. The most common activities are found in green investments and in optimizing production flows. The limitations of this study are that the market research was conducted between September and October 2024, and time may be a constraint. In future research, the open period for responses will be extended.

In using this data in questionnaire research, the authors received the consent of the respondents before they answered the questions asked.

Conflict of Interest

All authors declare that they have no conflicts of interest.

Author details

Larisa Ivascu[1*], Alin Artene[1] and Marius Pislaru[2]

1 Management, Politehnica University of Timisoara, Timisoara, Romania

2 Engineering and Management, Gheorghe Asachi Technical University of Iasi, Iasi, Romania

*Address all correspondence to: larisa.ivascu@upt.ro

References

[1] Fonseca LM, Domingues JP, Dima AM. Mapping the sustainable development goals relationships. Sustainability (Switzerland). 2020;**12** (8). DOI: 10.3390/SU12083359

[2] Bianchi F, Melosi L. Inflation as a fiscal limit. SSRN Electronic Journal. 2022. DOI: 10.2139/ssrn. 4205158

[3] Tuliakov V, Alekseievska H, Yakubovsky S. Legal and economic aspects of monetary regulation of the European system of central banks. InterEULawEast. 2021;**8**(2). DOI: 10.22598/iele.2021.8.2.4

[4] Wasan P, Kumar A, Luthra S. Green Finance Barriers and Solution Strategies for Emerging Economies: The Case of India. IEEE Transactions on Engineering Management. 2024; **71**:414–425. DOI: 10.1109/TEM.2021. 3123185

[5] Brühl V. Green finance in Europe — strategy, regulation and instruments. Intereconomics. 2021;**56**(6):323–330. DOI: 10.1007/s10272-021-1011-8

[6] Ruel-Bergeron S, Patel J, Kazi R, Burch C. The UN Sustainable Development Goals (SDGs). In: Ruel-Bergeron S, Patel J, Kazi R, Burch C, editors. Assessing Public Health Needs in a Lower Middle Income Country. 1st ed. Abingdon: Routledge; 2020. p. 9. DOI: 10.1201/9781003080220-8

[7] Birindelli G, Ferretti P, Ferri G, Savioli M. Regulatory reform and banking diversity: Reassessing Basel 3. Annals of Finance. 2022;**18**(4):429–456. DOI: 10.1007/s10436-021-00406-3

[8] Liu F, Stentoft L. Regulatory capital and incentives for risk model choice under Basel 3*. Journal of Financial Econometrics. 2021;**19**(1):53–96. DOI: 10.1093/jjfinec/nbaa029

[9] Feridun M. Green finance: Do innovation, fintech and financial transparency play a role? Applied Economics Letters. 2023;**32**:131–134. DOI: 10.1080/13504851.2023.2244227

[10] Meng X, Shaikh GM. Evaluating environmental, social, and governance criteria and green finance investment strategies using fuzzy AHP and fuzzy WASPAS. Sustainability (Switzerland). 2023;**15**(8). DOI: 10.3390/su15086786

[11] Karasmanaki E, Grigoroudis E, Galatsidas S, Tsantopoulos G. Citizen Satisfaction with Renewable Energy Investments: A Multi-Criteria Satisfaction Analysis. Energies. 2023;**16** (10):3979. DOI: 10.3390/en16103979

[12] Karasmanaki E, Grigoroudis E, Galatsidas S, Tsantopoulos G. Citizen satisfaction with renewable energy investments: A multi-criteria satisfaction analysis. Energies (Basel). 2023;**16**(10). DOI: 10.3390/en16103979

[13] Ivascu L. Green marketing strategies and market barriers to sustainability products design. In: Proceedings of 2018 the 8th International Workshop on Computer Science and Engineering, WCSE 2018. 2018. DOI: 10.18178/wcse.2018.06.138

[14] Dragan F, Ivascu L. Sustainable development: ICT, new directions, and strategies. Environmental Footprints and Eco-Design of Products and

Processes. 2021. DOI: 10.1007/978-981-16-4562-4_5

[15] Klepacki B, et al. Investments in renewable energy sources in basic units of local government in rural areas. Energies (Basel). 2021;**14**(11). DOI: 10.3390/en14113170

[16] Li Z, Kuo TH, Siao-Yun W, Vinh LT. Role of green finance, volatility and risk in promoting the investments in renewable energy resources in the post-covid-19. Resources Policy. 2022;**76**:102563. DOI: 10.1016/j.resourpol.2022.102563

[17] Ivascu L. Green marketing strategies and market barriers to sustainability products design. In: Proceedings of the 8th International Workshop on Computer Science and Engineering (WCSE 2018); 28–30 June 2018; Bangkok, Thailand: p. 834–838. DOI: 10.18178/wcse.2018.06.138

[18] Ivascu L, Sarfraz M, Domil A, Bogdan O, Burca V. Assessment of country institutional factor on sustainable energy target achievement in European Union. Economic Research-Ekonomska Istrazivanja. 2022;**35**(1). DOI: 10.1080/1331677X.2021.2002706

[19] Yasmeen R, Shah WUH, Ivascu L, Tao R, Sarfraz M. Energy crisis, firm productivity, political crisis, and sustainable growth of the textile industry: An emerging economy perspective. Sustainability (Switzerland). 2022;**14**(22). DOI: 10.3390/su142215112

[20] Abubakar M, Che Y, Ivascu L, Almasoudi FM, Jamil I. Performance analysis of energy production of large-scale solar plants based on artificial intelligence (machine learning) technique. Processes. 2022;**10**(9):1843. DOI: 10.3390/pr10091843

[21] Zhihong L, Ivascu L. Cloud computing resource dynamic optimization considering load energy balancing consumption. TELKOMNIKA (Telecommunication Computing Electronics and Control). 2016;**14** (2A):18. DOI: 10.12928/telkomnika.v14i2a.4344

[22] Draghici A, Ivascu L. Green Manufacturing in the Context of Circular Economy In: Draghici A, Ivascu L, editor. Sustainability and Innovation in Manufacturing Enterprises: Indicators, Models and Assessment for Industry 5.0. Singapore: Springer Singapore; 2022. p. 1–15. DOI: 10.1007/978-981-16-7365-8_1

[23] European Central Bank. Changes to the operational framework for implementing monetary policy [Internet]. Frankfurt am Main: European Central Bank; 2024 March 13. Available from: https://www.ecb.europa.eu/press/pr/date/2024/html/ecb.pr240313~807e240020.en.html [Accessed: 2024-March-15]

[24] The group of twenty (G20). Available from: https://www.g20.in/en/about-g20/about-g20.html [Accessed: 2024-March-10]

Section 3

Financing Risk and Resilience

Chapter 3

Disaster Risk Financing and Clean Energy Management in Sub-Saharan Africa: A Systematic Review and Comparative Cases (Malawi, Zimbabwe, Zambia)

Daniel Jemiard Mmasomwayera Sinkula

Abstract

The study examines the integration of disaster risk financing (DRF), clean energy management (CEM), and climate financing (CF) in Sub-Saharan Africa (SSA), with a focus on enhancing financial resilience against climate-related disasters in Malawi, Zimbabwe, and Zambia. DRF prepares financially for emergencies to protect livelihoods and assets. CEM is crucial for transitioning to a sustainable and climate-resilient economy in SSA. CF supports mitigation and adaptation projects that address climate change challenges in vulnerable communities, providing funding for disaster preparedness, response, and recovery efforts. The study employs a systematic review methodology to analyze literature on DRF, CEM, and CF in Malawi, Zimbabwe, and Zambia, focusing on trends, tools, and perceptions related to DRF and CEM. The synergic comparative analysis highlights the integrated strategies needed for DRF and CEM in Malawi, Zimbabwe, and Zambia. The study further examines the perceptions of non-profit organizations (NPOs) regarding DRF based on country-specific NPOs. In Malawi, NPOs report poor coordination between the government and international non-governmental organizations (INGOs), while in Zambia, NPOs emphasize the need for collaboration and innovative DRF strategies tailored to local conditions. Zimbabwean NPOs, on the contrary, struggle with limited funding and poor cooperation between sectors. The study recommends strengthening institutional capacities for successful DRF as well as the adoption of innovative financial instruments, such as green bonds, to optimize resource allocation in Malawi, Zimbabwe, Zambia, and the broader SSA region.

Keywords: DRF, risk layering, parametric insurance, ARC replica, Cat DDO, green bonds, mini-grids

1. Introduction

Disaster risk financing (DRF) involves preparing financially for emergencies to reduce the need for disruptive actions and protect livelihoods and assets. Non-profit organizations (NPOs, used interchangeably with NGOs herein, non-governmental organizations) in Sub-Saharan Africa (SSA) frequently address the regional disaster challenges. However, their service and the sustainability of humanitarian interventions depend on ready access to adequate financing arrangements. The recent increase in disaster frequency and scale, among other factors, highlights the growing importance of effective DRF for these organizations. This study explores the integration of DRF, clean energy management (CEM), and climate financing (CF) within NPOs in the SSA countries of Malawi, Zimbabwe, and Zambia. It highlights the critical role of DRF in enhancing financial resilience against climate-related disasters, which disproportionately affect these nations. The research identifies key strategies for improving DRF mechanisms, including mobilizing private capital, fostering public–private partnerships (PPPs), and implementing innovative financial instrument tools. By employing a systematic review methodology, the study analyzes existing literature that uncovers perceptions of DRF among NGOs and the barriers they face in integrating DRF into clean energy projects. While there is a recognition of the importance of integrating DRF into clean energy strategies, significant barriers exist, including insufficient funding, a lack of awareness, and regulatory frameworks and policies. The findings show the need for tailored financial frameworks that address the region's unique challenges, ultimately promoting sustainable development and resilience in the face of CEM and CF as they integrate with DRF for synergy. It also aligns DRF with CEM while emphasizing innovative financial instruments and collaborative approaches to effective disaster risk management (DRM) that align with equitable CF in the SSA region, mainly in Malawi, Zimbabwe, and Zambia.

2. Literature review

DRF is the intentional allocation of financial resources to allow for quick funding of disaster response and recovery operations, typically through different financial instruments, while CF attempts to raise funds to support mitigation and adaptation projects that address climate change challenges faced by underprivileged communities. On the contrary, CEM supports the transition to a sustainable and climate-resilient economy by promoting investment in clean energy (used interchangeably with renewable energy) and low-carbon technology. These financial management strategies have integrated procedures and objectives, making them relevant for disaster-prone countries in SSA, the region being one of the most susceptible to natural catastrophes in the world. The catastrophically induced impacts include infrastructure damage, power disruptions, environmental concerns, financial losses, lower agricultural productivity, food insecurity, and rising poverty. Because CEM helps to improve both disaster response and climate adaptation skills, DRF integrates with CEM and CF in assessing and minimizing the potential threats posed by disasters while managing and putting in place effective plans to lower potential risks and promote resilience and long-term strategies in response to disasters.

2.1 DRF, CEM, and CF

DRF, CEM, and CF seek to establish systems that support funding for preparedness, response, and recovery efforts, as well as prevention, as noted in the study by Linnerooth-Bayer and Hochrainer-Stigler [1], which highlights that risk financing can be used to complement and stimulate risk reduction. The goal is to ensure that funds are utilized transparently and efficiently. Evaluating and managing risks is crucial to building financial resilience, which involves identifying, minimizing, and transferring potential risks, as well as managing financial risks. Financial losses from disasters are increasing globally in less developed countries, prompting efforts to improve financial risk analysis, create new financial tools, and enhance cooperation.

2.1.1 Disaster risk financing

DRF in anticipatory action (DRFAA) is shifting its focus toward introducing early warning systems and proactive approaches, which allocate funds based on predictions and risk assessments rather than solely reacting to needs after a disaster has occurred. This approach aims to prevent or minimize the most severe effects of disasters by enabling swift action before they worsen [2]. Consequently, financial vulnerability assessment is taking the lead in disaster management of climate-related natural hazards, which have disproportionately affected countries with lower economic status and low-income populations. As in the study by Mechler et al. [3], many highly exposed developing countries, struggling with drained tax revenues, substantial debt, and insufficient international aid, have been unable to secure sufficient and timely funding to replace or repair damaged infrastructure and revive livelihoods following major disasters. However, over time, their DRF strategies have incorporated risk identification, risk assessments, risk acceptance, risk transfer, and risk monitoring and evaluation into national and local DRF plans. For example, the Organisation for Economic Co-operation and Development (OECD) has formulated guidelines to fund disaster risks and has staged international events to disseminate best DRF practices [4–6]. Another regional integration is carried out by the World Conference on Disaster Risk Reduction in Sendai, Japan, and the Sendai Framework for Disaster Risk Reduction 2015–2030. The framework serves as the primary guiding instrument for DRM within the context of sustainable development and poverty eradication [7] globally.

Financial management tools for DRF, CEM, and CF are important, as they help governments, vulnerable groups, and people recover after a disaster, and one example is alternative risk transfer arrangements. These include items such as cat bonds, whereby, when transferring its insurable risk, the entity may employ a swap, which is then classified as an insurance derivative [8]. Another option available is regional risk pools, whereby countries collaborate across international borders and split the costs of recovery. The four notable tools are CCRIF SPC (Caribbean Catastrophe Risk Insurance Facility Segregated Portfolio Company), serving the Caribbean and Central America. The other one is the Caribbean and Central America Regional Risk Pool SPC, which offers financial aid to governments in the Caribbean and Central America in the event of catastrophic hurricanes and earthquakes [9]. Other pooling tools are African Risk Capacity (ARC); the Pacific Island Countries Regional Risk Pool (PCRIC); and the Southeast Asia Disaster Risk Insurance Facility

(SEADRIF), covering ASEAN Plus Three (APT) cooperation countries, whose cooperation started in 1997 and was incorporated in 1999.

The ARC Group consists of the African Union Specialized Agency, established in 2012, and the Insurance Company Limited, a hybrid mutual insurer and the commercial affiliate of the group, founded in 2014. The ARC Group assists African governments in enhancing their capabilities, enabling them to craft more effective plans, prepare adequately, and respond efficiently to disasters caused by severe weather conditions, outbreaks, and epidemics. The ARC also establishes Pan-African climate response systems by employing innovative financial methods, including risk pooling and risk transfer, tailor-made for African nations susceptible to natural disasters [10, 11]. For example, the financial resources from ARC have strengthened Malawi's proactive approach to financing disaster risks, reducing reliance on costly post-disaster financing and increasing resilience. The Government of Malawi joined the second phase of the Regional Emergency Preparedness and Access to Inclusive Recovery (REPAIR) program, phase 2, in July 2025 to access a broader range of tools for managing disaster risks, joining Angola, Burundi, Seychelles, and Zambia. In addition to Malawi, Zimbabwe's participation in the ARC provides limited drought insurance coverage, with low cession rates. For example, Zimbabwe joined ARC in 2015 and took the policy in 2019 with inadequate coverage. Drought insurance was triggered at a one-in-four-year loss, covering about US$130 million, leading to cession rates that increased from 2% in 2019 to 9% in 2021 but dropped to 1.7% in 2022. Apart from country-based ARC [12], humanitarian partners also purchased ARC's Replica policy, covering over 1.2 million people since 2019.

The PCRIC is a regional catastrophe risk insurance system designed to help Pacific Island nations mitigate the financial consequences of natural disasters and climate change. The system ensures financial safety by combining risks and delivering coverage via insurance payouts during disasters. The system helps vulnerable nations mitigate the harsh consequences of events such as cyclones, floods, and earthquakes [13]. Similarly, the SEADRIF is a regional initiative that creates and implements disaster risk finance (DRF) and insurance solutions for Southeast Asian countries. It prioritizes enhancing financial resilience against natural disasters by providing access to risk transfer instruments such as insurance and reinsurance. Disaster risk finance strategies in the region heavily rely on transferring disaster-related costs and risks to another responsible party, compensating policyholders for losses following a disaster through financial instruments such as catastrophe bonds and swaps [14–16].

In Malawi, the DRF strategy aims to strengthen the financial capacity plan for disasters. It has recently developed a Social Cash Transfer Program (SCTP), as it involves better risk assessment as a means of creating a set of financial tools that help in DRF. The SCTP focuses on stronger institutions that seek financial safety through risk transfer, whereby, as noted by Malawi [17], incentives from the program encourage people to get insurance that will help in times of emergencies while providing financial protection for boosting financial safety from climate risks, which is ideal for small farmers in Malawi using a generic risk-layering approach.

Zimbabwe's DRF strategy aims to handle climate shocks and other disasters by assessing the existing systems and covering institutions, laws, and finances. Disasters include droughts, floods, and storms and are financed and managed by the national risk insurance through the African Development Bank via the Africa Disaster Risk Financing (ADRiFi) Capacity Building Project to enhance the country's skills. The

DRF strategy is part of the national plans, and it includes the National Development Strategy 1, which aligns with the sustainable development goals (SDGs). While the ADRiFi program works to improve financial safety for African nations, Zimbabwe is one of those countries, as past support from ADRiFi included payouts during the 2024/2025 El Niño drought in Zimbabwe [18].

Zambia's DRF strategy uses prearranged DRF tools for effective disaster response and recovery, earmarking for risk retention through budget contingencies, reserve funds, and risk transfer through micro insurance and sovereign insurance. Zambia also uses the protection gap, which measures the difference between disaster costs and the prearranged financing available to address them. The protection gap highlights the need for adequate prearranged finance to manage disaster risks effectively, since different financing instruments are suited for different types of disaster events, with risk retention being more cost-effective for frequent, low-impact events [19]. In addition, Chitondo et al. [19] noted that the country's disaster management framework and parties involved in disaster readiness, relief, and restoration efforts, while incorporating local expertise in minimizing the effects of disasters, created a set of funding tools. They also offer insurance and backup credit for disaster risk assessment, disaster response, and recovery costs. Zambia effectively managed disaster costs as a result of the World Bank's Disaster Risk Financing and Insurance (DRFI) Program's involvement. The program is consistent with Zambia's National DRM Framework of 2017 to 2030 and the global Sendai Framework, which spans from 2015 to 2030 [20].

2.1.2 Clean energy management

Managing clean energy involves many strategies, and the goal is to use renewable energy (clean energy is used interchangeably with renewable energy) fully, as it also aims to reduce harm to the environment. SSA needs clean energy technologies, as the region's energy demand is growing fast. SSA countries have many renewable energy resources, such as water, sun, and wind power. They can use these resources to meet their energy needs. However, one notable obstacle is getting enough funding, thus affecting renewable energy use. Investing in renewables depends on financial de-risking programs as a solution to make renewable energy technologies cheaper. For example, due to such technologies, solar power could provide 10% to 15% of Africa's electricity by 2050 [21]. Lower finance costs could significantly increase solar power's contribution; yet NGOs that specialize in enhancing CF in SSA are hindered by donor fatigue. Furthermore, while clean energy technologies and CF are closely linked, removing climate financial barriers is now critical.

More studies have been conducted on the importance of clean energy. This is especially true for SSA and its countries. The study by Pawar [22] points out that notable advances have been achieved in solar and wind energy, showing significant efficiency improvements (5% for solar photovoltaic (PV) and 10% for wind turbines) and decreases in costs (67% and 64%, respectively) over the last decade. Furthermore, it is observed that hydropower is the most efficient energy source, with a conversion rate of 90% for available energy, despite being limited by geographical factors.

Amidst the pressing need for sustainable development during climate change, the shift to renewable energy is crucial for reinforcing energy security and promoting economic resilience. This can only be realized by setting up an initial support fund to facilitate policy implementation, sustaining the current trajectory of rising innovation investment, and leveraging diversified market strategies, just like in China and beyond. This includes redefining the leadership role in the global energy transformation [23].

The study by Ang et al. [24] noted that renewable energy generates power that produces no greenhouse gas emissions from fossil fuel combustion and decreases certain types of air pollution. Thus, returning to sustainability is a suitable approach to address potential energy needs, combat rising temperatures, drive social and economic development, and reduce the impact of temperature change, environmental issues, and health concerns. However, their growth is limited by challenges to climate change mitigation, including market failures for clean energy, a lack of awareness, restricted access to raw materials required for future clean energy use, and daily carbon emissions [25]. Therefore, enhancing access to energy, energy usage, and consumption boosts energy liquidity, thereby minimizing the effects of climate change in less developed countries [26, 27].

In the SSA region, the deployment of clean energy projects is affected by various obstacles. These obstacles seriously hinder the advancement and growth of this industry. One of the main challenges is the lack of potential climate finances targeting renewable energy projects due to the reluctance of local banks to make investments in this high-risk sector [28]. These banks impose high lending interest rates on local NPOs, even though the government and other INGOs may show interest in subsidizing these renewable projects. In the quest for resilience, SSA's climate DRF focuses on adapting to climate change strategies in climate-smart agriculture. Thus, the importance of strengthening the relationship between the private and public sectors to improve capital allocation in projects and reduce financing costs while adopting innovative DRF approaches [29].

2.1.3 Climate financing

Climate-related DRM and transfer are heavily reliant on CF instruments. For instance, in national risk financing instruments, governments set up disaster reserve funds for swift relief and recovery efforts. The Philippines' Calamity Fund currently amounts to \$390 million, established for pre-disaster operations and recovery as of 2019. Before disasters strike, national risk financing instruments earmark budget reserves to cover losses. In response to flood alerts in 2019, Ethiopia's Flood Task Force allocated financial reserves, impacting more than 2.8 million people. Ex-post sources, such as budget reallocation and tax increases, tend to mobilize more slowly and should be deployed after calamity funds have been depleted [30]. While ex-post sourcing is important, Zambia's ability to access ex-post financing for disaster recovery is limited due to economic constraints.

International climate risk financing instruments receive vital backing from international sources, helping vulnerable countries with disaster recovery and enhancing their resilience. These comprise contingent credits and donor assistance; rapid financial access after a disaster is enabled by the World Bank's contingent credit lines. For example, the Seychelles was the first African country to gain access to the Development Policy Loan with the "Catastrophic Draw Down Option" or "Catastrophe Deferred Drawdown Option" (Cat DDO) in 2014 [31]. The other source is through CDRF (Community Disaster Resilience Fund), the term "CDRF" describes the use of strategic financial strategies to aid nations in managing the expenditures related to climate-related events and natural catastrophes. Malawi was among the first in the region to implement a national CDRF strategy and has made significant strides since the program started in 2019. These instruments involve mechanisms that help manage climate risks by shifting financial burdens to external parties. Examples of these arrangements include catastrophe bonds and contingent debt facilities. Catastrophe bonds are financial instruments that transfer specified risks to investors, with payments initiated by qualifying disasters. The Climate Vulnerable Forum (CVF) suggested implementing a Contingent Convertible Finance (CCF) initiative that would provide grant support, which would turn into concessional debt for successful projects, and conversely, debt could be converted into grants for failing initiatives [32].

Community-level climate risk transfer involves implementing vital community-based risk management strategies to address climate vulnerabilities nationwide. Climate risks can be better addressed when understood and assessed by the communities. For this reason, the communities need to be aware of the current and future climate risks to evaluate their ability to respond, which requires using scientific climate models and satellite data to make accurate predictions. It is no wonder that development practitioners play a key role in strengthening community resilience against climate risks. Similarly, development practitioners advocate for capacity development and risk-sharing initiatives by raising awareness about climate risks and available risk management options to support informed community decision-making.

3. Methodology

Recognizing the need for specialized approaches to data collection, the study methodology followed the PRISMA framework (Preferred Reporting Items for Systematic Reviews and Meta-Analyses) [33–35]. This method guarantees transparency, reproducibility, and thoroughness in locating and examining literature. The data collection was related to DRF, CF, and CEM from various regions with a global comparison, while focusing on the SSA countries of Malawi, Zimbabwe, and Zambia. These countries were chosen purposefully due to their diverse geographical, ecological, socio-economic, and political settings. They were one country during the Federation of Rhodesia and Nyasaland, which was formed in 1953 and ended in 1963 [36]. These countries were selected to reflect diverse levels of susceptibility to DRF and clean energy project implementations, disaster frequency and severity, and data availability for adaptation and CEM initiatives.

To ascertain the extent to which this literature explicitly addresses DRF in CEM: Perceptions of Sub-Saharan Region's NPOs, the data search, selection, and evaluation were based on the availability of numerous articles and studies from literature and gray literature drawn from various internet-based platforms. These platforms include Google Scholar, Scopus, Web of Science (WoS), ResearchGate, and Search Engine Optimization (SEO). Keywords such as "climate energy," "DRF," "NGOs," "perception in DRF," "DRF," and "climate financing" yielded numerous results from significant articles. The search was refined by combining these strings with "AND" and "OR" Boolean operators. For example, "DRF" AND "clean energy management" AND "Sub-Saharan Africa" OR "world" OR "Malawi" OR "Zimbabwe" OR "Zambia" OR "Caribbean" OR "Central America" OR "Seychelles"

OR “Southeast Asia” OR “Pacific Island nations” OR “China” AND “NGOs” “ perception” OR “Disaster Risk Reduction Frameworks.”

The literature review process from the searched and selected publications involved a methodical evaluation of the acquired data to determine its accuracy, completeness, and appropriateness for analysis, with an emphasis on its relevance. Sources of literature for review were obtained through the use of the PRISMA approach to reduce bias and increase its practical applicability across a wide range of DRF in clean energy, with relevant application to climate change in SSA nations. The author subsequently selected 304 published articles based on relevance to the study, from which 85 articles were removed before screening, as they were repetitive. The 186 publications (see **Figure 1**) were screened to select the primary studies directly related to the objective of this study, its regional focus, and Sub-Saharan countries. Subsequently, 57 publications were removed because a review of their abstracts showed they were not eligible for the systematic review, as they were duplicates. Furthermore, 23 publications were ineligible, and 106 publications were selected for analysis.

The systematic review procedures were based on trends in DRF, DRF tools, CEM overview, synergy among DRF, CEM, and CF; perceptions of the SSA region’s NPOs; and challenges in DRF in SSA. The publications used highlight and complement the literature on the impacts of climate change on the clean energy industry in SSA; some older papers and materials that were considered significant were also included.

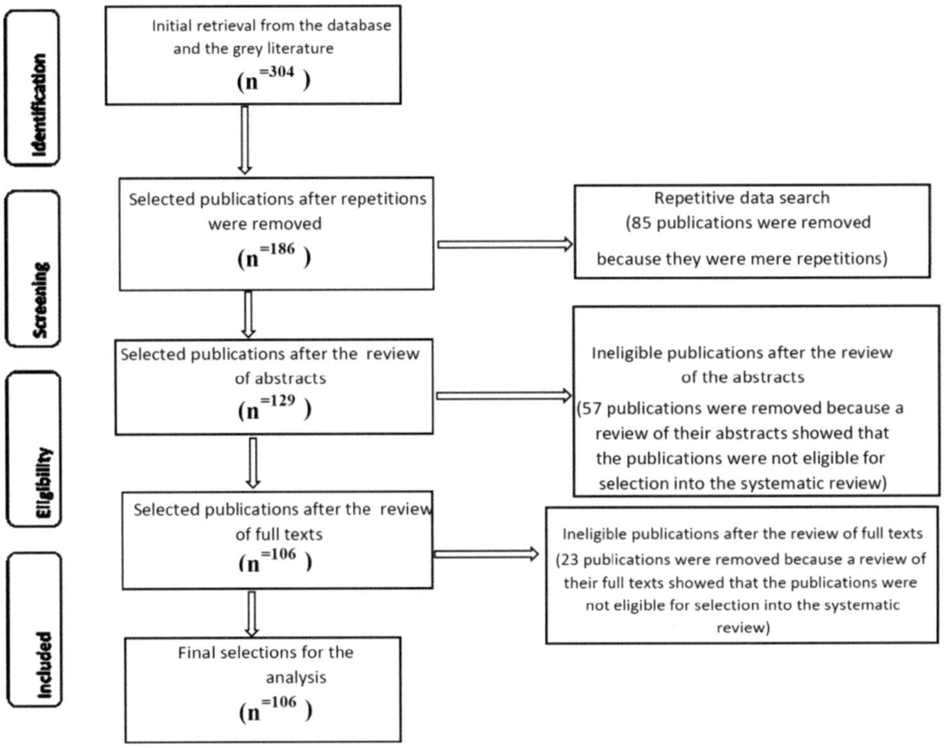

Figure 1.
PRISMA flowchart – source author’s own design.

To facilitate a comprehensive analysis, the extracted data were grouped into key themes for analytical purposes, focusing on DRF and CEM in three SSA countries. Examinations of studies were conducted to pinpoint patterns in the types, frequencies, and intensities of climate-related energy disasters, such as floods, droughts, and extreme temperatures. This was supplemented by presentations on trends in DRF, highlighting the increased focus on prearranged DRF, the importance of anticipatory action, growing awareness of disaster risk financial vulnerability, and the integration of DRF into climate energy and CF, all tailored to address each country's distinctive challenges and best practices where applicable. A backward and forward citation tracking method, proposed by Hu et al. [37], was utilized to identify supplementary studies missed in the initial search for data.

The methodological process, as outlined in the PRISMA flowchart (see **Figure 1**), demonstrates the systematic selection and rejection of studies, thereby guaranteeing the reliability and thoroughness of the review. This approach enabled a detailed analysis of the current state of knowledge on major disaster financing instruments, clean energy vulnerability, the impacts of climate finance, and the strategies used for adaptation and DRF across the three selected SSA countries.

4. Synergy among DRF, CEM, and CF: A conceptual framework

DRF, CEM, and CF integrate opportunities for mutually beneficial outcomes. DRF seeks to enhance financial preparedness against natural disasters and the effects of climate change. It involves creating strategies to secure funding in advance for relief, recovery, and reconstruction efforts, potentially utilizing tools such as insurance and risk transfer mechanisms. Clean energy aims to decrease greenhouse gas emissions through the development and implementation of renewable energy sources and energy efficiency initiatives. CF includes financial resources from various providers to aid in efforts for climate change mitigation (such as clean energy projects) and adaptation strategies.

The synergy among DRF, CEM, and CF fosters cooperation between the public and private sectors, aids climate-vulnerable regions, and strengthens risk management by producing more accurate data. Strengthening the connection between climate change adaptation and disaster risk reduction is essential for effective risk management. The synergy promotes more effective policies and investments that address existing and future risks and enhances technical knowledge and expertise in managing both clean energy and non-clean energy systems. The increasing impacts of CEM necessitate a unified approach to adaptation and risk reduction, as both are crucial for building resilience in communities and economies, reducing risks, and enhancing adaptive capacity. Therefore, the opportunity to implement renewable energy (renewable energy used interchangeably with clean energy) projects while mitigating climate change and natural disaster risks in the sub-Saharan region is facilitated by combining DRF, CEM, and CF initiatives.

CEM requires investments in renewable energy, energy efficiency measures, and sustainable energy infrastructure, ultimately leading to lower exposure to climate change impacts, reduced greenhouse gas emissions, improved air quality, wider energy access for communities, and a decrease in vulnerability. Furthermore, CEM regulates and optimizes the production, distribution, and usage of clean energy resources. Well-managed clean energy resources boost economic well-being and deliver vital services to their citizens via effective energy planning and management systems. To address escalating energy security requirements and bridge the disparity between energy consumption and production, a wide range of renewable energy technologies is developed and deployed in developed and developing countries [38]. The extent to which clean energy systems can withstand major catastrophes remains uncertain, despite combining insights from CEM and disaster risk reduction management presenting intriguing opportunities to tackle this challenge [39].

4.1 Interconnections among CF, CEM, and DRF

Figure 2 illustrates the interrelationship among DRF, CEM, and CF, which are interrelated foundations for fostering climate and disaster resilience. Arrows demonstrate how they support each other: CF provides funding for clean energy initiatives to mitigate climate threats; DRF addresses expenses from unavoidable disasters while encouraging risk reduction measures; and CEM decreases dependence on fossil fuels and bolsters community resilience, all aimed at achieving sustainable development objectives and alleviating poverty. The overarching objective that unifies all three pillars is to develop adaptive capacity in communities and systems to deal with the effects of climate change and natural disasters. This enables climate-resilient development by providing funding for CEM, early warning systems, disaster preparedness, and nature-based solution.

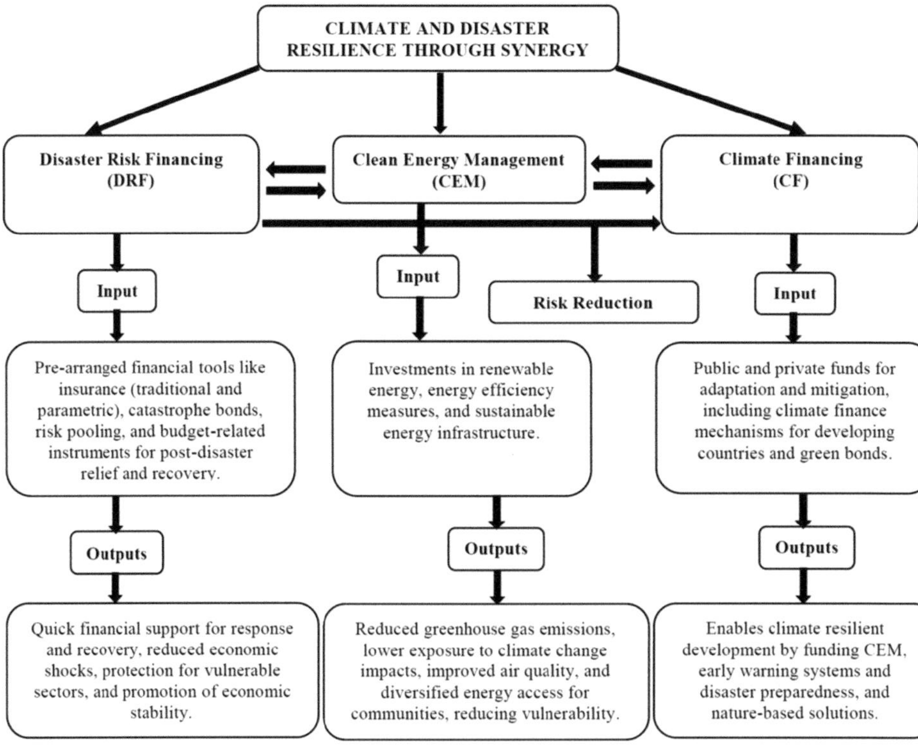

Figure 2.
Conceptual framework on synergy among DRF, clean energy management, and climate financing. Source: author's own design.

In **Figure 2**, furthermore, the arrow from CF to CEM indicates that climate finance allocates funds to clean energy projects that mitigate climate risks. The arrows from CEM to DRF and to CF represent a diversified, sustainable energy system that reduces the vulnerability of communities to climate and disaster shocks, thus lowering the need for DRF and enhancing the effectiveness of CF. An arrow from DRF to CEM or CF indicates that financial protection mechanisms can establish a stable setting, allowing for long-term investments in CEM and CF through the creation of a safety net and the facilitation of risk transfer. Arrows from CF and DRF to risk reduction demonstrate that both climate and disaster financing support risk assessment, early warning systems, upgraded infrastructure, and capacity building, which, in turn, decrease future risk exposure. In summary, the arrows illustrate interactions with risk factors, including poverty, unsustainable land use, and governance issues, which affect the effectiveness of synergy in achieving successful integration.

4.2 Financial instruments

Various financial tools are now accessible to support climate energy, as the financing of CF, CEM, and DRF projects has undergone substantial changes. Particularly in the Sub-Saharan region, financial tools are essential for DRF in clean energy projects. Venture capital and private equity investments target financing high-risk, high-reward renewable energy projects. Investors who focus on impact investing, especially those in the renewable energy sector, offer equity funding to renewable energy projects that demonstrate potential in implementing cutting-edge energy solutions. These investments are usually long-term in nature, allowing clean energy projects to develop and expand clean energy technologies over time.

Several financial instruments can be employed to decrease risks and enhance the attractiveness of renewable energy projects. Such initiatives include green bonds, sustainability-linked loans, decentralized finance, and crowdfunding. Decentralized finance is defined in the study by Hysaj [40] as financial services provided on a public blockchain, accessible globally via the internet. Green bonds are financial instruments used to fund environmentally sustainable projects linked to climate change mitigation. Green bonds have also become a promising means to finance the shift toward a more flexible, lower-carbon economy [41]. In addition, green bonds are a crucial element of a holistic strategy employed by investors in renewable energy to align financial markets with climate goals and meet global greenhouse gas reduction targets [42]. The sustainability-linked loan is another financial instrument gaining popularity. Unlike traditional loans, which are based on a borrower's strength, sustainability-linked loans are linked to the borrower's performance in meeting environmental, social, or governance objectives. These loans for initiatives involving renewable energy or clean energy offer flexible terms and the opportunity for reduced interest rates if specific sustainability criteria are met.

Sub-Saharan nations can improve their capacity to implement renewable technologies by prioritizing the reduction of financial risks. In the study by Mirgen and Tepeli [43], it is noted that energy usage is increasing and becoming a significant concern to the environment as it determines living standards and drives economic growth. The shift toward sustainable energy depends on access to sustainable energy sources, the usage of advantageous energy carriers, the improvement of energy efficiency, and the reduction of the long-term environmental impacts of energy systems.

5. Comparative case analyses

According to the case analysis in **Table 1**, the perceptions from Malawi, Zimbabwe, and Zambia show how climate resilience necessitates integrated strategies in DRF, CEM, and CF.

5.1 Comparative case analysis based on DRF

Table 1 discusses the challenges in DRF across Zimbabwe, Malawi, and Zambia, particularly in response to climate-induced disasters. In Zimbabwe, Cyclone Idai in 2019 revealed significant shortcomings in institutional structures, prompting calls for improved disaster legislation and proactive risk management [44, 45]. Malawi's strategy (2019–2024) incorporates risk-layering tools and a shock-responsive SCTP to mitigate reliance on external aid during droughts and floods [48]. Meanwhile, Zambia faces increasing climate vulnerability impacting agriculture and essential services, with initiatives from international organizations addressing the financial protection gap. These assessments are essential for developing comprehensive national strategies that connect climate change, energy volatility, and financial risks [51].

5.2 Comparative case analysis based on CEM

Table 1 shows that CEM in Zimbabwe, Malawi, and Zambia faces significant challenges, particularly in rural areas where access to consistent electricity is limited. In Zimbabwe, the Renewable Energy Fund has allocated $1 million for a solar power plant in Dairibord, aimed at reducing carbon emissions while supporting local dairy farmers and schools. The UN Capital Development Fund has contributed $810,000 for a solar project at Mater Dei Hospital, with an additional $6 million for a solar facility in Guruve District to enhance local businesses and social services [46]. In Malawi, energy access remains constrained, particularly in rural communities. UNICEF has initiated programs using solar power to electrify clinics and boreholes, while community energy systems have introduced energy kiosks and mini-grids, improving access for low-income households and enhancing healthcare outcomes [49]. In Zambia, despite abundant renewable resources, many citizens lack electricity and clean cooking options, rendering the country vulnerable to climate-related droughts. The Zambia National Energy Compact aims for universal clean energy access by 2030 by promoting distributed renewable energy, particularly solar power, in rural regions to reduce dependence on hydroelectric sources and enhance resilience to climate change impacts [52].

5.3 Comparative case analysis based on CF

Table 1 shows that severe drought conditions in Zimbabwe have led to a significant 50% decrease in cereal yields compared to the previous five-year average. In response, the Gungano Fund has emerged as an effective climate finance model at the community level, integrating local savings with donor funds to foster sustainable livelihoods and improve living conditions in slums. This initiative has enhanced climate resilience through installations like dry toilets in flood-prone areas, solar energy systems, and damage repair following extreme weather events. The

Country	Case study area	Key challenge	Implemented strategy or project	Results and lessons learned
Zimbabwe	Disaster Risk Financing	Inadequate institutional and legal structures, along with a passive approach to disaster response – highlighted by Cyclone Idai in 2019 – resulted in significant fiscal and social expenses.	After Cyclone Idai, evaluations were conducted to gauge the nation's financial readiness, revealing considerable institutional and policy deficiencies. The assessment advocated for enhanced disaster legislation and improved governance [44, 45].	The lessons emphasized the importance of strengthening laws related to disaster risk, enhancing coordination systems, and developing skills to transition from reactive to proactive disaster management.
	Clean Energy Management	Rural areas face difficulties in obtaining consistent electricity, relying on the national grid, which experiences underproduction and incurs significant expenses for expansion.	The Renewable Energy Fund (REF Zimbabwe) has allocated $1 million for a solar power plant at Dairibord Zimbabwe, aiming to reduce carbon emissions by 26,000 tons annually. The facility will support 480 small dairy farmers and two schools in Chipinge. The UN Capital Development Fund has also provided $810,000 for a solar project at Mater Dei Hospital in Bulawayo. An additional $6 million will be directed toward a solar facility in Guruve District [46].	Enhanced local businesses, improved social services, and showcased a framework for developing self-sufficient, off-grid "eco-communities."
	Climate Financing	Zimbabwe faced severe droughts, causing a 50% drop in cereal yields compared to the average of the past five years.	The Gungano Fund, a local initiative, offers a strong model for climate finance at the community level by combining community savings with donor funds to support sustainable livelihoods and improve slums [47].	Projects funded by Gungano improve climate resilience through installing dry toilets in flood-prone areas, using solar energy systems, and repairing damage after extreme weather. The fund adapted to Zimbabwe's unstable economy by changing from cash repayments to repayments in building materials during hyperinflation.
Malawi	Disaster Risk Financing	Disasters, particularly droughts and floods, cause severe fiscal pressure and macroeconomic instability, with the government historically relying on unpredictable external aid.	Malawi's Disaster Risk Financing Strategy (2019–2024) Implemented a risk-layering framework that integrates pre-arranged (ex-ante) tools, such as contingency funds and parametric insurance, alongside post-disaster (ex-post) strategies [48].	The strategy aims to secure prompt financing by establishing pre-arranged financial access and minimizing dependence on external aid and budget adjustments. The shock-responsive social cash transfer program (SCTP) has effectively expanded to support vulnerable families affected by droughts.

Country	Case study area	Key challenge	Implemented strategy or project	Results and lessons learned
	Clean Energy Management	Limited energy access, especially in rural, hard-to-reach communities, hinders access to basic services like healthcare.	UNICEF Renewable Energy Solutions: Focuses on using solar power to provide electricity to clinics and boreholes in rural areas. Community Energy Systems: Implementation of energy kiosks and mini-grids has provided feasible energy options for low-income households [49].	Improved health outcomes by ensuring vaccine cold chains and enabling water pumps. These systems have proven that community-managed energy systems are feasible and can foster a transition from traditional energy sources.
	Climate Financing	Challenges include low insurance coverage, with just 5% of adults having insurance, and a limited insurance market that makes it difficult to manage climate risks and for insurers to make profits [50].	To help with climate finance, efforts are underway to complete carbon market frameworks and build institutional capacity, with support from organizations like the UNDP.	The country's limited fiscal space necessitates additional financing through grants, highly concessional public financing, and private sector inflows.
Zambia	Disaster Risk Financing	The region experiences a high level of susceptibility to climate change, with regular occurrences of floods and droughts affecting essential sectors such as agriculture, energy, and water, which result in a notable "protection gap."	Zambia's strategy for financing climate and disaster risks involves efforts from organizations such as the Potsdam Institute for Climate Impact Research (PIK) and the German Corporation for International Cooperation (GIZ), which focus on measuring the financial "protection gap" associated with the effects of floods and droughts. This assessment serves as a crucial basis for formulating a thorough national strategy [51].	Recognized the particular financial requirements and measured the discrepancy between anticipated disaster effects and current financial protection, establishing a foundation for a stronger national strategy. This underscores the connection between climate change, hydro-energy volatility, and financial risk.
	Clean Energy Management	Although Zambia has plenty of renewable resources, a significant number of its citizens do not have access to electricity or clean cooking options, and the heavy dependence on hydropower makes the country susceptible to droughts caused by climate change.	Zambia National Energy Compact: This initiative is in line with SDG 7 and aims for universal access to clean energy by 2030, with a focus on distributed renewable energy (DRE), such as solar power, in rural regions to reduce dependency on hydroelectric sources [52].	Fast-tracking the adoption of off-grid solar and clean cooking solutions via creative financing strategies. Expanding the energy mix is essential for enhancing resilience to the effects of climate change.
	Climate Financing	Zambia has faced serious droughts recently, causing cereal yields to drop by 43% compared to the average over the last five years. The country has high poverty levels, with 54% of its people living below the poverty line, which restricts their ability to invest in adaptation measures [53].	There is a proposal for an Adaptation Fund project aimed at building joint adaptation systems for smallholder farmers and communities in areas near Transfrontier Conservation Areas (TFCAs).	Climate change adaptation strategies should be intensified, targeting small holder farmers and communities near transfrontier areas.

Source1: Author's own design.

Table 1.
Perceptions of case studies in DRF and clean energy.

fund has adapted to Zimbabwe's fluctuating economy by shifting from cash repayments to repayments in building materials amid hyperinflation [47]. Meanwhile, Malawi's CF landscape is hindered by low insurance coverage – only 5% of adults are insured – and a limited insurance market that complicates risk management for insurers [50]. Efforts are being made to develop carbon market frameworks and bolster institutional capacity, with backing from organizations such as the UNDP. However, the country's constrained fiscal space necessitates further financing through grants and targeted private sector investments. Similarly, Zambia has recently experienced severe droughts, resulting in a 43% decline in cereal yields and high poverty levels, with 54% of the population living below the poverty line, restricting their capacity to invest in adaptation initiatives [54]. Drawing from prior experiences, Zambia has proposed an adaptation fund project aimed at establishing joint adaptation systems for smallholder farmers and communities near Transfrontier Conservation Areas (TFCA). Consequently, intensified climate change adaptation strategies should focus on supporting smallholder farmers and communities situated in these critical areas.

6. Perceptions of Sub-Saharan region's NPOs on DRF

NPOs, used interchangeably with non-governmental organizations, dedicate more resources to disaster management than governmental and intergovernmental agencies, providing expertise and knowledge development on all facets of disaster. The distinction between governmental, intergovernmental, and non-governmental agencies implicitly influences the perspectives of various DRF in CEM in SSA. NPO perceptions in DRF derive from different ideas and views held by one side about the other. In SSA, NPOs frequently consider disaster relief funding to be essential; however, they face challenges such as inadequate DRF, insufficient funding opportunities, and a lack of capacity-building programs for their staff. These challenges influence how NPOs view DRF, given that nations in this area are especially vulnerable to various severe natural disasters like droughts and floods [54]. SSA's NPOs perceive DRF as a means to enhance the effectiveness of humanitarian assistance; however, advancement is obstructed by insufficient support from national and local governments, along with the lack of comprehensive legal and policy frameworks that protect NPOs' efforts to fulfill their mandate in obtaining reliable and disaster-specific funding. Consequently, NPOs perceive the need to intensify DRF during and after the occurrence of disasters in the SSA. Similarly, governments in SSA nations use a range of ex-ante (composed of risk transfer, risk retention, and external risk finance) and ex-post financing (composed of risk retention and external risk finance drawn from grants, loans, and other external finances) to deal with disasters, despite their heavy dependence on external help. In recent years, SSA countries have experienced a significant amount of innovation in new mechanisms for providing contingent coverage of disaster-related losses. For example, Kita [50] states that Malawi introduced disaster insurance, and it is increasingly recognized as a vital tool for addressing climate-induced loss and damage in Malawi; however, its adoption remains low due to various challenges.

Additionally, NPOs perceive DRF as a strategy for improving the efficiency and effectiveness of delivering humanitarian aid and disaster recovery initiatives. Moreover, NPOs acknowledge an increasing demand for more advanced, proactive financing solutions that go beyond the reactive nature of DRF, which is frequently inadequate. Thus, Chikoto-Schultz et al. [55] note that NPOs make a big difference in the community. They also contribute significantly to the underrepresented, marginalized, and vulnerable populations by providing services, advocating for them, developing their leadership and expressive skills, fostering community and democracy, and driving innovation.

6.1 Perceptions cases of Malawi, Zambia, and Zimbabwe

NPOs are proactively preparing communities to face future challenges by combining insurance and financial support mechanisms, going beyond simply responding to crises. NPOs are, therefore, essential to promoting resilience and sustainability in the face of rising disaster risks in the area by expanding access to financial resources and promoting inclusive disaster funding. For example, in the study by Pauw et al. [56], climate-related disasters are resulting in increasingly severe economic losses; consequently, such disasters may lead to widespread poverty and hinder the development goals of numerous nations. The challenges in question imply that achieving the SDGs in SSA may heavily depend on how disaster management policies enhance hydrological resilience and climate change adaptation. For example, in Zimbabwe, following Cyclone Idai, assessments of the country's fiscal preparedness were carried out, uncovering significant shortcomings in both institutions and policies (see **Table 1**). Meanwhile, opportunities exist for the Paris Agreement and Nationally Determined Contributions (NDCs), as noted in literature [57–59], to boost climate finance disbursements to Africa. In the end, NPOs' influence on DRF systems will rely on their ability to innovate, work together, and effectively advocate for those in need as their role continues to change.

6.1.1 Perceptions of Malawi

Malawi faces significant humanitarian crises primarily due to extreme weather events, climate change, cyclones, and tropical storms. The Malawi government leads humanitarian responses through the Department of Disaster Management Affairs and DRM Committees. As such, Malawi receives support from international NGOs (INGOs) that play crucial roles in disaster management, particularly during flood emergencies, and these include, but are not limited to, UN Women, World Vision, Action Aid, and WFP. These INGOs are actively involved in disaster responses, focusing on resource provision and agenda setting. NPOs perceive weak collaboration and coordination among INGOs when dealing with DRF, climate change, and CEM management. In Malawi, disaster responses suffer from poor coordination between the government and INGOs, despite the United Nations Office for the Coordination of Humanitarian Affairs (UN OCHA) responding to disasters through the cluster system approach. However, local NPOs (LNPOs) and local non-governmental organizations (LNGOs) do not have enough adequately trained and experienced staff dedicated to the communication process with the clusters [60]. On the contrary, many NGOs prepare contingency plans without proper input from DoDMA, leading to poorly informed strategies. NGOs are concerned about the commitment to funding long-term disaster risk reduction initiatives. Additionally, the judiciary's involvement in disaster management plans is regarded as vital, although it frequently feels detached from the actual circumstances [61].

Many local NPO networks, such as civil society organizations, collaborate with INGOs in Malawi. Others include faith-based organizations (FBOs), such as Churches Action in Relief and Development, which provide humanitarian assistance but often operate in isolation and lack formal connections to climate change project initiatives. Despite support from INGOs, the Malawi government still suffers from the imposition by INGOs, as they dominate in the control of disaster financial resources. This dominance frequently ignores LNPOs' distinct advantages and local expertise. Moreover, DRF-related projects fail to meet the actual needs of LNGOs and LNPOs in terms of capacity-building initiatives in DRF [62]. In this case, DRF in Malawi is perceived as the most significant challenge in the non-profit-making sector due to concerns about corruption, which negatively impact donor trust in LNGOs, while they perceive that DRF sources are limited.

NPO perceptions are crucial in deciding how to plan, prepare, mitigate, and respond to disasters at any time in DRM and CEM. General trends show that NGOs in Malawi contribute to DRM through initiatives in disaster prevention, preparedness, and recovery. For instance, although Malawi receives assistance from NPOs that are important in implementing community-focused flood risk management initiatives, a thematic analysis in the study by Smith [63] found that donors typically prefer established technological methods rather than local expertise. This restricts their integration into projects. However, organizations in Malawi think that the local community's grasp of project planning and implementation does not align with their needs or priorities. On the contrary, there is continued NPO donor dependency in Malawi through a begging bowl pattern from donors. In the study by Sharma et al. [64], it is noted that in Malawi, donors contribute 55% of health spending, yet the aid system is fragmented with many funding sources and implementing partners, which causes poor coordination between government goals and donor efforts. To help them plan strategically and create successful financial protection programs, NPOs in Malawi perceive that having better access to and utilization of information on DRF leads to effective disaster preparedness. Thus, in the study by Dewa et al. [65], evidence from the flooding situation in Nsanje District in Malawi shows that Malawi has improved its DRM institutional structures. This has been possible due to shifting its policies from reactive to proactive in dealing with disasters.

6.1.2 Perceptions of Zambia

The larger problems of financial viability, transparency, and the national-level regulatory framework for DRM influence the perceptions of Zambian NPOs regarding DRF. Given Zambia's rising vulnerability to climate-related disasters, it is critical to increase its involvement in DRM by strengthening cooperation with the government and foreign partners. For instance, Kachumi et al. [66] demonstrates that the Church Health Association (CHA), which receives support from international funders, has significantly impacted the lives of vulnerable communities, particularly in rural regions. Due to inadequate funding and economic challenges, NPOs in Zambia fail to maintain financial stability. Additional issues include domestic concerns related to dependence on donors and a dearth of long-term strategy. Using DRF techniques to enhance the sustainability of NPOs, on the contrary, is one of the best methods for managing disaster risk finance. Budgeting and cash flow management are just two examples of effective DRF strategies that are necessary for long-term stability.

NPOs in Zambia are well aware that a key barrier is the scarcity of resources for DRR technologies and their implementation at the national and local government levels. The need for increased collaboration and coordination between NPOs, government agencies, and the commercial sector to create and implement integrated DRF programs is evident. According to NPOs, there is a scarcity of innovative, effective, and flexible DRF strategies that prioritize risk and are tailored to local conditions to transfer climate risk insurance and other mechanisms to protect vulnerable communities. Public perception highlights the shortcomings of local governments, as well as the inability of stakeholders, such as NPOs, to take advantage of and benefit from opportunities to incorporate climate change adaptation and DRR into development initiatives and to make efficient use of DRF tools. International partners and NPOs have discovered a significant gap in risk awareness that limits the effectiveness of current DRF methods. Data-driven methods are creating financial protection measures, and the evolution of climate change is providing solutions to mitigate these effects and safeguard communities at risk.

As demonstrated by the World Bank-supported projects, NPOs in Zambia are actively engaged in helping government initiatives to improve the national DRM framework through technical support and capacity development [67]. Organizations like the World Food Programme (WFP) promote inclusive risk financing that is integrated with larger social protection programs to increase community resilience in Zambia, while at the same time promoting a risk layering strategy to manage risks of different magnitudes. This strategy involves integrating risk transfer (such as insurance-linked securities, insurance, and reinsurance) with risk retention (such as contingency funds and anticipatory actions that include contingent credit and reserves) [68]. On the contrary, PPP in Zambia is seen by NPOs as essential for fostering favorable market conditions for sustainable insurance and for integrating DRF instruments with private sector investments. However, in the study by Clarke and Mahul [68], the PPP Act in Zambia had not made a considerable impact on the private sector's choice to engage in PPP projects.

6.1.3 Perceptions of Zimbabwe

Zimbabwean NPOs respond to various disasters caused by changes in climate change that subsequently affect clean energy resource projects. However, these NPOs perceive that disaster risk finance is hindered by insufficient access to funding resources. Additionally, NPOs also perceive that there is a lack of coordination between the public and private sectors in disaster aid preparedness, response, and recovery initiatives. Similarly, in many cases, perceptions emphasize the poor cooperation between NPOs, government organizations, and the private sector, which hinders resource sharing and the development of all-inclusive financial solutions. The environment in which NPOs function during relocations caused by disasters at Chingwizi presents several obstacles. Some of these obstacles arise internally within the NPOs, while others are external factors. The government of Zimbabwe has been criticized for limiting the operational space for NPOs that aim to support individuals impacted by disaster-related relocations. NPOs are required to funnel all their aid through the office of the Provincial Affairs Minister instead of engaging directly with the people they intend to help [69].

Because NPOs often have small budgets, they struggle to allocate financial resources toward DRF to address the rising frequency and severity of disasters. Risk transfer methods and insurance, which are necessary for mitigating the financial consequences of disasters, are difficult to obtain for NPOs due to their lack of competence in DRF and disaster risk finance. Many NPOs require extensive training to implement successful DRF projects. Alternatively, in the study by Bongo et al. [70], Zimbabwe's framework for disaster risk reduction and its execution have not yet encompassed some of the most marginalized and vulnerable groups, especially the terminally ill, people with disabilities, and the extremely impoverished. A top-down method to disaster management is frequently criticized for hindering resilience and compromising preparedness among the parts of society most impacted by disasters. Furthermore, disaster risk reduction efforts have generally been tailored to address the needs and priorities of able-bodied individuals, largely neglecting those with different disabilities. The need for the development of low-cost, innovative financial tools like micro-insurance or community-based risk funding programs is a crucial insight.

7. Findings from the study

Obstacles that have previously prevented the efficient funding of climate change adaptation projects form the basis toward gaining a thorough understanding of these findings.

7.1 Lack of uniform policies

The lack of uniform policies that consider the diverse socioeconomic conditions of various nations in SSA is one of the main challenges [71]. Insufficient funding streams, resulting from fragmented approaches to climate finance, frequently fail to meet the pressing needs for adaptation. Prior errors have stemmed from poor local capacity, regulatory framework gaps, and the lack of uniform resource mobilization policies, which would make it easier to provide a unified voice in support of uniform policy changes at regional and national levels.

7.2 Lack of adaptable DRF strategies in climate change

DRF is now an important component of SSA's climate adaptation plans. As in some studies [72, 73], firms are launching their climate change initiatives, presenting a pressing concern for corporate governance as the board monitors and evaluates these efforts within their organizations. Governments and development partners frequently operate independently, paying little attention to the interdependence of climate threats and possible mitigation measures. The insurance systems that may foster proactive planning and investment in climate resilience are undermined by this lack of comprehensive and systematic methodologies. Even when insurance may be required, a lack of robust stakeholder participation and communication limits its effectiveness as a tool for moving toward sustainable climate adaptation. In some cases, obstacles such as poor infrastructure and limited access to trustworthy climate data hinder investments in disaster risk reduction strategies. This issue has been exacerbated by the ongoing repercussions of climate change, which have raised questions about the efficacy of current financial legislation and systems.

7.3 Lack of climate adaptation techniques

Finding best practices that meet local needs and empower vulnerable communities has become crucial in the context of climate energy in SSA when it comes to effective DRF. As climate change continues to shift at an unprecedented rate, it is best to develop and implement effective adaptation strategies that enable vulnerable communities to become more resilient and cope with the consequences. The most successful climate adaptation techniques used in vulnerable communities in SSA draw on the most current research on climate adaptation strategies [74, 75].

8. Recommendations

Several recommendations come to light in addressing the inconsistent policies surrounding DRF, clean energy, and CF in SSA. The initiatives involve promoting regional cooperation, boosting institutional capabilities, and refining policy and regulatory structures. Moreover, implementing innovative financial tools such as green bonds and risk-sharing mechanisms, leveraging private sector investments, and increasing transparency are crucial for establishing trust and optimizing fund deployment. It is, therefore, essential for stakeholders to address the funding shortfall and enhance the resilience of vulnerable groups more effectively.

Countries in SSA encounter substantial difficulties in creating adaptable strategies for DRF in response to climate change. The main reasons for this are limited financial resources, insufficient institutional capabilities, and a scarcity of conventional public funding. Combating this requires a comprehensive approach, including investments in climate adaptation to reduce disaster relief costs, an expansion of international grants and concessional loans, and the involvement of the private sector. It also requires the development of new financial tools designed to boost fiscal resilience in light of increasing climate-related disaster frequencies.

Among the most effective adaptation techniques to climate change are early warning systems and strategies that combine climate data, agricultural services, and healthcare services. Additionally, strategies are needed to improve coordination among government agencies, development partners, and financial institutions. An assessment of financial challenges, obstacles, and potential solutions should correspond with innovative funding methods tailored to the specific situation, rather than applying universal approaches both before and after disasters occur. In order to find lasting solutions for managing and funding the risk of disasters in SSA's changing climate–energy environment, it is essential to involve all stakeholders at every level, from national governments to local communities.

9. Conclusion

In conclusion, the integration of DRF with CF and CEM in SSA, particularly in Malawi, Zimbabwe, and Zambia, enhances financial resilience in the face of escalating climate-related disasters. The systematic review of existing literature reveals that while NGOs are increasingly recognizing the importance of incorporating DRF into CEM initiatives, significant barriers still hinder this process, including limited funding, insufficient public awareness, and inadequate regulatory frameworks. This study underscores the pressing need for innovative financial instruments and the promotion of PPPs to mobilize private capital more effectively. Furthermore, an emphasis on tailored financial systems in addressing the unique challenges of SSA region countries facilitates the long-term growth and resilience of communities against the adverse impacts of disasters. Furthermore, the necessity for a holistic approach that ensures the harmonious alignment of DRF with CEM and CF is paramount. This encompasses comprehensive financial management strategies aimed at overseeing prevention, preparedness, response, and recovery efforts. Thus, the facilitation of transparency and efficiency in DRM is contingent upon the establishment of effective financial mechanisms to direct financial resources toward sustainable development objectives. Vulnerable communities are better equipped to confront the obstacles presented by disasters and climate change. As such, the study lays the groundwork for further research in DRF to enhance synergy among DRM practices, CEM, and CF, and to address perceptions among NPOs in DRF.

Conflict of Interest

The author declares no conflict of interest.

Acknowledgements

The authors acknowledge the use of Grammarly for language editing in the chapter.

Author details

Daniel Jemiard Mmasomwayera Sinkula
Nkhoma University, Lilongwe, Malawi

*Address all correspondence to: danieljemiardmmasomwayera@gmail.com

References

[1] Linnerooth-Bayer J, Hochrainer-Stigler S. Financial instruments for disaster risk management and climate change adaptation. Climatic Change. 2015;**133**(1):85–100

[2] Uddin S, Jannat M. Anticipate to Protect: Prioritizing Health and a Proposed Framework for Anticipatory Action in a Changing Climate. 10.13140/RG.2.2.15901.42725. 2025

[3] Mechler R, Hochrainer S, Pflug GC, Lotsch A, Williges K. Assessing the financial vulnerability to climate-related natural hazards. World Bank Policy Research Working Paper, (5232). 2010

[4] Baubion C. OECD risk management: Strategic crisis management. 2013

[5] Ermolieva T, Ermoliev Y. Catastrophic risk management: Flood and seismic risks case studies. In: Wallace SW, Ziemba ST, editors. Applications of Stochastic Programming. Laxenburg, Austria: Society for Industrial and Applied Mathematics; 2005. p. 425–444

[6] Michel-Kerjan E, Hochrainer-Stigler S, Kunreuther H, Linnerooth-Bayer J, Mechler R, Muir-Wood R, Ranger N, Vaziri P, Young M. Catastrophe risk models for evaluating disaster risk reduction investments in developing countries. Risk Analysis. 2013;**33**(6):984–999

[7] Surianto S, Alim S, Nindrea RD, Trisnantoro L. Regional policy for disaster risk management in developing countries within the Sendai framework: A systematic review. Open Access Macedonian Journal of Medical Sciences. 2019;7(13):2213

[8] Sibindi AB. The art of alternative risk transfer methods of insurance. Risk Governance and Control: Financial Markets and Institutions. 2015;**5**(4):216–226

[9] CCRIF SPC Caribbean and Central American Partnership for Catastrophe Risk Insurance: Pooling Risk to Safeguard against Catastrophes Generated by Natural Events. [report]. 2014. World Bank approach using copulas. Environmetrics. 2012; 23(8):663-72. [Accessed: 2025-July-24]

[10] Fernández Lopera CC, Mendes JM, Barata EJ. The differential risk transfer: A new approach for reducing vulnerability to climate-related hazards. Disaster Prevention and Management: An International Journal. 2022;**31**(5):550–564

[11] World Food Program. (2018). ARC Replica: WFP's Partnership with the African Risk Capacity (ARC) for the Expansion of Climate Risk Insurance. Rome. Available from: https://docs.wfp.Org/api/documents/WFP-0000063792/download/?_ga=2.106353091.1310370720.1527488109-1613948681.1525264650 [Accessed: 2025-August-07]

[12] Mhlanga G, Chirisa I. The future of managing disasters and risks in Zimbabwe and beyond. In: Kasimba R, Chirisa I, editors. Disaster Risks and Hazards: Practice for Resilience-Case for Zimbabwe, Singapore: Springer Nature Singapore; 2025:285–298

[13] Jain DK, Chida A, Pathak RD, Jha R, Russell S. Climate risk insurance

in Pacific Small Island Developing States: Possibilities, challenges and vulnerabilities—A comprehensive review. Mitigation and Adaptation Strategies for Global Change. 2022;**27** (3):26

[14] Jha AK, Stanton-Geddes Z, editors. Strong, Safe, and Resilient: A Strategic Policy Guide for Disaster Risk Management in East Asia and the Pacific. Washington, DC: World Bank Publications; 2013

[15] Radu D. Disaster Risk Financing: Limiting the Fiscal Cost of Climate-related Disasters. Luxembourg: Publications Office of the European Union; 2022 Nov

[16] Nobanee H, Nghiem XH. Climate catastrophe insurance for climate change: What do we know and what lies ahead? Current Opinion in Environmental Sustainability. 2024;**66**:101–395

[17] Malawi Social Cash Transfer Program (SCTP). Providing Support to the Less Privileged. Available from: https://www.unicef.org/malawi/reports/malawi-social-cash-transfer-program-sctp-longitudinal-impact-evaluation-0 [Accessed: 2025-August-4]

[18] Malawi, Zimbabwe Receive Financial Boost for Climate Change Mitigation Programs. The Herald. October 15, 2024. Available from: www.herald.co.zw

[19] Chitondo L, Chanda CT, Mpolomoka DL, Ngulube L. Disaster management and mitigation strategies in Zambia: A systematic review. World Journal of Advanced Research and Reviews. 2024;**21**(3):2403–2420

[20] Government of Zambia. Zambia National Disaster Risk Management Framework (2017-2030): Operationalising the Sendai Framework. Available from: https://www.preventionweb.net/publication/zambia-national-disaster-risk-management-framework-2017-2030 [Accessed: 2025-August-02]

[21] Sweerts B, Dalla Longa F, van der Zwaan B. Financial de-risking to unlock Africa's renewable energy potential. Renewable and Sustainable Energy Reviews. 2019;**102**:75–82

[22] Pawar S. Harnessing the power of renewable energy: A study of sustainable sources and technologies. Research Journal of Applied Sciences. 2024;**3**:163–169

[23] Pourasl HH, Barenji RV, Khojastehnezhad VM. Solar energy status in the world: A comprehensive review. Energy Reports. 2023;**10**:3474–3493. DOI: 10.1016/j.egyr.2023.10.022

[24] Ang T-Z, Salem M, Kamarol M, Das H, Nazari M, Natarajan P. A comprehensive study of renewable energy sources: Classifications, challenges, and suggestions. Energy Strategy Reviews. 2022;**43**.100939. DOI: 10.1016/j.esr.2022.100939

[25] Ndayishimiye V, Zhang X, Nibagwire D, Simiyu P, Dushimimana G, Bikorimana S. Environmental benefits of modern power system and clean energy. In: E3S Web of Conferences 2019 (Vol. **107**, p. 02006). EDP Sciences

[26] Joon R. Renewable energy sources: A review. In Journal of Physics: Conference Series. Vol. **1979**. Bristol: IOP Publishing; 2021. p. 012023. No. 1

[27] Jaber JO, Badran OO, Abu-Shikhah N. Sustainable energy and environmental impact: Role of renewables as clean and secure source of energy for the 21st century in Jordan. Clean Technologies and Environmental Policy. 2004;**6**(3):174–186

[28] Adeosun T, Amosu C, Omitogun O, Amusa O, Morenikeji B. Renewable energy for sustainable development in developing countries: Benefits to the environment. Journal of Energy Research and Reviews. 2023;**13**:1–25. DOI: 10.9734/jenrr/2023/v13i3262

[29] Michoud B, Hafner M. Financing Clean Energy Access in sub-Saharan Africa: Risk Mitigation Strategies and Innovative Financing Structures. Cham: Springer Nature; 2021

[30] Onabowale O. The Impact of Green Bonds on Energy Efficiency: Financing Clean Technology and Energy Infrastructure. 2024

[31] Weber R, Musshoff O. Risk-contingent credit for sovereign disaster risk finance. International Journal of Disaster Risk Reduction. 2021;**56**:102–105

[32] Jie H, Khan I, Alharthi M, MW Z, Saeed A. Sustainable energy policy, socio-economic development, and ecological footprint: The economic significance of natural resources, population growth, and industrial development. Utilities Policy. 2023;**81** (101490). DOI: 10.1016/j.jup.2023.101490

[33] Parums DV. Review articles, systematic reviews, meta-analyses, and the updated preferred reporting items for systematic reviews and meta-analyses (PRISMA) 2020 guidelines. Medical Science Monitor: International Medical Journal of Experimental and Clinical Research. 2021;**27**:e934475–1

[34] Moher D, Clarke M, Ghersi D, Liberati A, Petticrew M, Shekelle P, LA S. Preferred reporting items for systematic review and meta-analysis protocols (PRISMA-P) 2015: Elaboration and explanation. Bmj. 2015;**349**:1

[35] Pauw WP, Castro P, Pickering J, Bhasin S. Conditional nationally determined contributions in the Paris Agreement: Foothold for equity or Achilles heel? Climate Policy. 2020;**20** (4):468–484

[36] Marmon B. Settler worldmaking: Reconfiguring the federation of Rhodesia and Nyasaland, 1953–62. Itinerario. 2023;**47**(1):91–109

[37] Hu X, Rousseau R, Chen J. On the definition of forward and backward citation generations. Journal of Informetrics. 2011;**5**(1):27–36

[38] Munene MB, Gerger Swartling Å, Thomalla F. Adaptive governance as a catalyst for transforming the relationship between development and disaster risk through the sendai framework? International Journal of Disaster Risk Reduction. 2018;**28**: 653–663

[39] Wen J, Wan C, Ye Q, Yan J, al E. Disaster Risk Reduction, Climate Change Adaptation and Their Linkages with Sustainable Development over the Past 30 Years: A Review. 2023. Available from: ncbi.nlm.nih.gov [Accessed: 2025-July-28]

[40] Hysaj A. Decentralized Finance. International Journal of Advanced

Natural Sciences and Engineering Researches. 2023;7(10):158–162

[41] Daszyńska-Żygadło K, Słoński T, Ligus M. Investment in renewable energy technologies from the perspective of Polish venture capital Funds. In European Financial Systems 2016. Proceedings of the 13th International Scientific Conference 2016 Jun 27 (pp. 79–85)

[42] Ozili PK. Decentralized finance research and developments around the world. Journal of Banking and Financial Technology. 2022;**6**(2):117–133

[43] Mirgen Ç, Tepeli Y. Effect of energy consumption on green bond issuance. Statistika: Statistics & Economy Journal. 2024;**104**(2):203–211

[44] Chatiza K. Cyclone Idai in Zimbabwe: An analysis of policy implications for post-disaster institutional development to strengthen disaster risk management. Oxfam. Retrieved from: https://policycommons.net/artifacts/1822168/cyclone-idai-in-zimbabwe/2560538/ [Accessed: 2025-September-12]

[45] Munsaka E, Mudavanhu C, Sakala L, Manjeru P, Matsvange D. When disaster risk management systems fail: The case of Cyclone Idai in Chimanimani District, Zimbabwe. International Journal of Disaster Risk Science. 2021;**12**(5):689–699

[46] UNCDF. Zimbabwe's First Renewable Energy Fund: Building a Distributed Energy Market from the Ground Up. Zimbabwe: UNCDF; Old Mutual; Joint SDG Fund; 2025

[47] Coger T, Dinshaw A, Tye S, Kratzer B, Aung MT, Cunningham E, Ramkissoon C, Gupta S, Bodrud-Doza M, Karamallis A, Mbewe S. Locally Led Adaptation: From Principles to Practice. Washington, DC, USA: World Resources Institute; 2022

[48] Krutham. Scaling insurance for climate resilience in Africa: Insights from Malawi [Internet]. African Climate Foundation; 2024 Dec 2. Available from: https://africanclimatefoundation.org/research-article/scaling-insurance-for-climate-resilience-in-africa/scaling-insurance-for-climate-resilience-in-africa-insights-from-malawi/ [Accessed: 2025-July-22]

[49] UNICEF Malawi and Partners. Sustainable Energy for UNICEF Malawi: Renewable energy solutions improve children's health and well-being in Malawi. 2021

[50] Kita SM. disaster insurance for climate loss and damage: Assessing challenges and opportunities to adoption in Malawi. Risk, Hazards & Crisis in Public Policy. 2025;**16**(1): e70002

[51] Potsdam Institute for Climate Impact Research (PIK) and GIZ. Climate and disaster risk financing in Zambia – A protection gap analysis. Available from: PreventionWeb.net. 2023

[52] Amoah M. Bridging the energy access divide: A policy gap analysis of 12 African national energy compacts under Mission 300 2025

[53] Thurlow J, Zhu T, Diao X. The impact of climate variability and change on economic growth and poverty in Zambia. 2009

[54] Hirsch T. Climate Risk Financing: A Brief Analysis of Financial Coping Instruments and Approaches to Close

the Protection Gap', Analysis Paper, No. 87, Brot für die Welt-Evangelisches Werk für Diakonie und Entwicklung e. V., S. 2019; 23

[55] Chikoto-Schultz GL, Xiao Y, Manson P, Amiri M. Non-profit sector organizational actions on risk reduction practices, policymaking participation, community and social contributions, and recovery. In Oxford Research Encyclopedia of Natural Hazard Science [Accessed: 2019-August-28]

[56] Pauw K, Thurlow J, Bachu M, Van Seventer DE. The economic costs of extreme weather events: A hydrometeorological CGE analysis for Malawi. Environment and Development Economics. 2011;**16**(2):177–198

[57] Granderson AA. Making sense of climate change risks and responses at the community level: A cultural-political lens. Climate Risk Management. 2014;**3**:55–64

[58] Willitts-King B, Weingärtner L, Pichon F, Spencer A. Risk-informed approaches to humanitarian funding: Using risk finance tools to strengthen resilience. ODI Report; 2020.

[59] Chirambo D. Mitigating disaster risks and vulnerabilities through climate finance and sustainable water management: Policy considerations for Sub-Saharan Africa and Malawi. In: Islam R, Uddin MM, Islam R, Babu MdA. editors. Disaster Risk Reduction for Resilience: Climate Change and Disaster Risk Adaptation. Cham Springer International Publishing; 2023. p. 117–139

[60] Stumpenhorst M, Stumpenhorst R, Razum O. The UN OCHA cluster approach: Gaps between theory and practice. Journal of Public Health. 2011;**19**. DOI: 10.1007/s10389-011-0417-3

[61] Manda S, Thindwa TTM. This one caught us unaware': Disaster politics and institutions during Cyclone Freddy in Malawi. International Journal of Disaster Risk Reduction. 2025;**117**:105144

[62] Helmold M, Samara W. PM in NGO and NPO. In: Helmold M, Samara W, editors. Progress in Performance Management: Industry Insights and Case Studies on Principles, Application Tools, and Practice; Jul 14 2019; Cham: Springer International Publishing. p. 149–154

[63] Smith J. Extreme weather jeopardises Malawi's development goals: Can the country achieve its socioeconomic goals amid the damage of its brutal cyclone cycle? The Conversation. 2025; Published online. [Accessed: 2025-September-10]

[64] Sharma L, Heung S, Twea P, Yoon I, Nyondo J, Laviwa D, Kasinje K, Connolly E, Nkhoma D, Chindamba M, Tebeje MT. Donor coordination to support universal health coverage in Malawi. Health Policy and Planning. 2024;**39**(Suppl 1):i118

[65] Dewa O, Makoka D, Ayo-Yusuf OA. Assessing capacity and implementation status of the disaster risk management strategy for health and community disaster resilience in Malawi. International Journal of Disaster Risk Science. 2021;**12**(5): 673–688

[66] Kachumi M, Nawa M, Mwanakatwe J, Menda D, Sichinga K, Shaikh J. (2025). Financial Sustainability of Local NGOs Funded by the Churches Health Association

of Zambia in Zambia. A Mixed Methods Survey. 10.21203/rs.3.rs-5822319/v1.

[67] Vaughan A, Hillier D. Ensuring Impact: The Role of Civil Society Organisations in Strengthening World Bank Disaster Risk Financing. London: Centre for Disaster Protection; 2019 Mar

[68] Clarke D, Mahul O. Disaster risk financing and contingent credit: A dynamic analysis. World Bank Policy Research Working Paper. (5693). [Accessed: 2011-June-01]

[69] Tarisayi K. Some Contemporary Challenges Faced by NGOs in Disaster Induced Relocations at Chingwizi Transit Camp, Zimbabwe. Zimbabwe. [Accessed: 2014-July-04]

[70] Bongo PP, Chipangura P, Sithole M, Moyo F. A rights-based analysis of disaster risk reduction framework in Zimbabwe and its implications for policy and practice. Jàmbá: Journal of Disaster Risk Studies. 2013;**5**(2):1–1

[71] Tippmann E, Ambos TC, Del Giudice M, Monaghan S, Ringov D. Scale-ups and scaling in an international business context. Journal of World Business. 2023;**581**:101397

[72] Surminski S, Barnes J, Vincent K. Can insurance catalyse government planning on climate? Emergent evidence from Sub-Saharan Africa. World Development. 2022;**153**:105830

[73] Africa Disaster Risk Management Team. Report on the Status of Disaster Risk Reduction in sub-Saharan Africa. World Bank, 2010

[74] Grove H, Clouse M, Xu T. Climate change risk: Challenge for corporate governance. Journal of Governance and Regulation. 2021;**10**(2):258–267

[75] Otundo Richard M. 21st Century Climate Adaptation Strategies for Vulnerable Communities in Sub-saharan Africa; an in-depth Literature Review. Available at SSRN 4823867. 2024 May 10

Section 4

Sectoral Pathways for Climate Solutions

Chapter 4

Coordinating Transportation and Land Use Strategies Across Scales: Lessons from California's Climate Action Planning Experience

Serena E. Alexander and Ahoura Zandiatashbar

Abstract

This chapter explores how regional and local governments can coordinate transportation and land use strategies to reduce greenhouse gas (GHG) emissions, drawing on lessons from California's Senate Bill 375 and its multilevel climate planning framework. It asks: To what extent do regional sustainable communities strategies (SCSs) and local climate action plans (CAPs) align in their approaches to reducing vehicle travel, and what does this alignment – or lack thereof – mean for climate action? Using qualitative content analysis of 25 planning documents from five diverse California metropolitan regions, the study examines 24 strategies across four categories: transportation infrastructure, land-use policies, transportation demand management, and cross-cutting issues such as collaboration and equity. The findings reveal that regional plans tend to emphasize modeling and best practices, while local plans are often more tailored to community needs and, in some cases, more ambitious. Strategies requiring significant infrastructure investment, such as transit or active transportation networks, benefit most from vertical integration, whereas locally driven innovation fills gaps through policies such as parking reform, urban greening, and community-centered engagement. Key barriers to coordination include mismatched authority, uneven technical capacity, and differing timelines, though cities with long-standing climate commitments often act as leaders. California's experience underscores that effective decarbonization requires both coordination across levels of government and room for local innovation. These lessons are transferable to other jurisdictions seeking to align transportation and land use planning in support of climate action.

Keywords: land use–transportation coordination, climate action planning, multilevel governance, greenhouse gas emissions mitigation, low-carbon development, vehicle miles traveled reduction, policy alignment, planning capacity, sustainable communities strategies

1. Introduction

California is often seen as a pioneer in climate policy, but even in this ambitious state, the road to lowering greenhouse gas emissions – especially from cars and trucks – is not easy. One of the most significant attempts to tackle this challenge came in 2008 with the passage of Senate Bill 375 (SB 375), also known as the Sustainable Communities and Climate Protection Act. The bill aimed to reduce greenhouse gas emissions from cars and light trucks – the state's largest single source – by reshaping how communities grow and how people move within them.

SB 375 was designed to bridge the gap between climate goals and urban development. It tasked each of California's Metropolitan Planning Organizations (MPOs) – the regional bodies responsible for long-range transportation investments – with developing a sustainable communities strategy (SCS). These strategies were supposed to integrate land use, housing, and transportation planning in ways that support emissions reduction targets set by the California Air Resources Board (CARB). The hope was that if people lived closer to jobs, services, and transit, they would drive less – and emissions would fall. **Figure 1** shows how climate planning strategies are linked between state, regional, and local governments in California.

From its inception, SB 375 was both visionary and challenging. As Brent Schoradt [1] noted shortly after the law passed, the success of the bill would depend heavily on the quality and ambition of these SCSs [1]. But as Elisa Barbour [2] later observed in her evaluation of early SCSs, translating high-level sustainability goals into effective plans and measurable outcomes proved more difficult than anticipated [2]. One issue was the potential disconnect between regional planning and local implementation.

While SCSs are mandatory under SB 375, Climate Action Plans (CAPs) – which many cities adopt to guide their local climate actions – are voluntary. This creates a potentially fragmented landscape in which local jurisdictions may or may not align their plans with regional strategies. Researchers like Frick et al. [3] have highlighted the tension between regional coordination and local autonomy, especially around issues like equity, housing, and growth [3]. Their work, along with that of Garde and

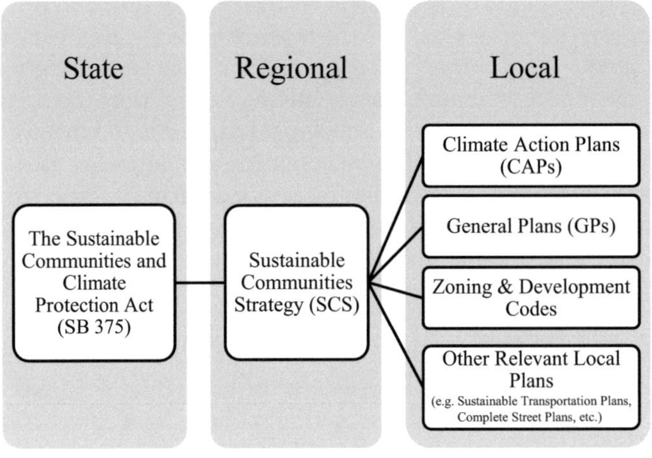

Figure 1.
Conceptual diagram linking climate planning at various levels of government in California.

Kim [4] on zoning reform [4], underscores how essential it is for local governments to not only participate in but also adapt regional strategies in ways that reflect their own capacities and priorities.

Research to date offers limited and, at times, conflicting evidence on whether regional plans effectively spur local voluntary adoption of sustainable transportation and land-use practices. A longitudinal analysis of plans in the Denver metropolitan area suggests that, while local development continues to be strongly influenced by market dynamics and local preferences, MPOs have had some success in advancing mixed-use and higher-density development that supports reductions in VMT and GHG emissions [5]. By contrast, a study of the Sacramento region showed that neighborhoods planned under the regional "Blueprint" principles did not attract most new residential growth; in fact, the areas rated most favorably experienced less development after the plan's adoption [6]. More recently, evidence from California indicates that local activist leadership – through initiatives such as parking policy reforms and community outreach – can outperform regional plan alignment in promoting nonauto commuting and advancing GHG reduction goals [7].

Moreover, there is a lack of evidence on how regional GHG reduction strategies are incorporated into local plans. A 2017 report from the National Institute for Transportation and Communities (NITC) analyzed MPO-developed regional transportation plans in California and Oregon and found that the establishment of regional GHG reduction targets led to greater attention to climate change in planning [8]. However, this report did not evaluate the influence of state mandates or regional plans on local planning documents. Another study highlighted the importance of systematically assessing how MPOs across California encourage local governments to pursue more compact, less car-dependent development to meet statewide GHG goals [9]. Specifically, the research suggests that MPOs typically conduct limited and inconsistent evaluations of how local land-use practices and built environment changes align with SB 375 and their corresponding regional SCS. This underscores the need for a comprehensive assessment of the alignment between regional SCS and local CAPs throughout California. Lessons learned would benefit other jurisdictions pursuing climate planning to address transportation emissions.

This chapter contributes to the literature by filling that gap. Using a comparative content analysis of 25 planning documents from five diverse California regions, we examine the alignment between regional SCSs and local CAPs. This cross-jurisdictional approach builds on prior case studies by offering a broader perspective, allowing us to identify patterns of strategy adoption, highlight areas of strong alignment and divergence, and explore how these patterns relate to decarbonization outcomes. In doing so, our work adds empirical depth to ongoing discussions in climate planning research.

This analysis focuses on 24 types of strategies, grouped into four categories:

1. Transportation infrastructure – investments and design choices that expand safe, low-carbon travel options, such as bike lanes, sidewalks, or EV charging stations.

2. Land-use policies – tools to shape where and how development occurs, including zoning changes that promote infill housing or mixed-use neighborhoods.

3. Transportation demand management (TDM) – efforts to reduce the need to drive by changing travel behavior, such as flexible work programs, transit incentives, or outreach campaigns.

4. Cross-cutting strategies – approaches that integrate equity, public engagement, or intergovernmental collaboration into planning and implementation.

Together, these strategy types reflect a comprehensive framework for transportation decarbonization – but only if they are effectively coordinated. Through this analysis, we found that while some cities and regions are closely aligned, others take their own path. That's not necessarily a bad thing – local innovation can thrive even without regional mandates – but a lack of coordination can also mean missed opportunities to reduce emissions more effectively.

Our research builds on earlier work and new resources from the CARB's Sustainable Communities and Climate Protection Program [10], which continues to track implementation challenges and lessons from SB 375. What follows is a deeper look into these patterns, peppered with real-world examples that show the promise – and the complexity – of planning for climate action across levels of government.

While this chapter focuses on California, the insights are relevant far beyond the state's borders. Around the world, cities and regions are grappling with how to align ambitious climate targets with on-the-ground decisions about land use and mobility. Whether it's Paris rethinking car access, Bogotá expanding bus rapid transit, or Canadian provinces integrating land use into climate planning, the challenge is the same: how to get multiple levels of government working toward a shared vision. California's experience offers both cautionary lessons and hopeful strategies. For any jurisdiction seeking to decarbonize its transportation sector while managing growth and equity, understanding what worked – and what didn't – in California's complex planning landscape can be an invaluable guide.

2. Methodological framework

This chapter draws on the qualitative portion of a broader study that examined the relationship between regional and local climate action planning in California [11]. The analysis focused on the transportation and land use strategies included in regional SCSs and local CAPs to understand the extent to which regional policy priorities are reflected in local plans.

2.1 Case selection

The study examined climate action planning across five major MPOs in California:

1. MTC/ABAG (Bay Area)

2. SCAG (Southern California)

3. SACOG (Sacramento area)

4. SANDAG (San Diego)

5. SJCOG (San Joaquin Valley)

These MPOs were selected to capture the breadth of California's geographic, demographic, and economic diversity, as well as the variety of institutional contexts in which Senate Bill 375 implementation occurs. Collectively, they encompass the majority of the state's population and transportation-related greenhouse gas emissions, making them critical arenas for understanding climate action planning in practice.

Within each MPO region, five cities were selected to capture the diversity of local contexts, including large and small jurisdictions, urban cores and suburban municipalities, and varying levels of governance capacity and climate planning experience. Selection criteria considered both population size and geographic distribution, ensuring inclusion of the largest cities as well as mid-sized and smaller municipalities, with locations dispersed across each region. Four cities in each region had adopted a CAP at the time of data collection, while one city had not, providing a point of comparison to assess the absence of formal climate action planning. This sampling approach allowed the analysis to encompass both high-capacity cities with established climate programs and those at earlier or preliminary stages of climate action.

2.2 Strategy framework

The strategies assessed in this study were identified through a systematic content analysis of the selected SCSs and CAPs. Drawing from the climate planning literature, iterative review of the plans, and targeted keyword searches, we developed a list of 24 strategies grouped into the aforementioned four categories.

These strategies were selected because they are widely recognized as policy levers for reducing transportation-related greenhouse gas emissions. For each plan, the presence or absence of each strategy was recorded, with a value of "1" assigned if the strategy was explicitly included and "0" if it was not. Strategies were counted as present if they appeared as a long-range public policy or program in the plan, regardless of whether they were legislative, incentive-based, or voluntary in nature.

To ensure clarity and consistency, two researchers pilot-tested the framework by independently coding one CAP from each metro region, as well as its relevant SCS. This process allowed them to refine definitions, resolve ambiguities, and establish coding rules. Results of the pilot test were compared to assess intercoder reliability, and discrepancies were discussed until consensus was reached. Once consistency was confirmed, the refined framework was applied to the full set of 25 cities and 5 MPOs. This two-step approach – framework development followed by pilot testing – ensured the reliability and validity of the qualitative content analysis.

2.3 Analytical approach

The qualitative analysis focused on identifying patterns, similarities, and differences between regional and local plans. This included examining:

- Which strategies appeared consistently across both levels of planning?

- Where local plans diverged from regional priorities.

- Which strategies were unique to local contexts or to regional frameworks?
- How the absence of a CAP influenced the scope of transportation and land-use strategies at the local level.

While this chapter predominantly presents the qualitative results of the plan analysis, descriptive comparisons across the sample are used to highlight notable trends. In some instances, examples from individual MPOs or cities are provided to illustrate broader findings, such as the vertical "flow down" of strategies from regional to local plans or the emergence of strategies that originate locally and are later reflected in regional planning.

3. Findings

3.1 Transportation strategies: Where plans align and why it matters

Let's start with the basics: how people move. **Table 1** summarizes the adoption of key transportation infrastructure and built environment strategies across regional SCSs and local CAPs, showing both the number and share of plans incorporating each strategy.

As depicted in **Table 1**, almost every region and city included strategies to make walking and biking safer and more attractive. "Active transportation," as planners call it, has become a widely accepted part of climate planning. Long Beach, for example, adopted a comprehensive Complete Streets Master Plan that doesn't just mention bike lanes – it weaves them into the city's identity. The plan ties together pedestrian safety, greenery, stormwater management, and access for people with disabilities, shaping capital projects across the city.

Strategy	SCSs adopting (n = 5)	% of SCSs	CAPs adopting (n = 20)	% of CAPs
Bicycle infrastructure	5	100%	19	95%
Pedestrian infrastructure	5	100%	19	95%
Complete streets	5	100%	12	60%
Mass transit expansion	5	100%	16	80%
Electric vehicle infrastructure	4	80%	18	90%
Ride-sharing/carpool programs	4	80%	15	75%
Low-carbon/alternative fuel vehicles	3	60%	18	90%
Autonomous vehicle infrastructure	2	40%	2	10%
Climate-friendly infrastructure	1	20%	14	70%
Vehicle idling reduction	2	40%	7	35%
Goods movement/freight policies	3	60%	10	50%

Table 1.
Adoption of transportation infrastructure/built environment strategies across MPOs and cities in California.

This reflects broader trends: all analyzed SCSs include active transportation strategies, and nearly all CAPs (19 of 20 cities) incorporate pedestrian and bicycle infrastructure. Complete streets policies are explicitly referenced in 10 CAPs, while several others defer to General Plans or specialized Active Transportation Plans. These strategies are reinforced by decades of state policy, including Caltrans Deputy Directives dating back to 2001 and culminating in the 2021 "Complete Streets" DP-37, which requires all Caltrans-funded projects to accommodate safe and comfortable conditions for all modes.

Mass transit is another central strategy. Every regional plan and 15 CAPs discuss transit improvements, though the level of specificity varies. Oakland's CAP, for instance, emphasizes equity by targeting underserved neighborhoods with faster buses, better stops, and shelters designed for extreme heat. Major urban hubs – Los Angeles, San Francisco–Oakland, and San Jose – plan to enhance system efficiency, reduce travel time, and improve inclusiveness, while midsize and smaller cities often supplement regional transit agency efforts with local infrastructure, technological tools like GPS tracking, and incentives for riders. Across plans, the alignment of local and regional transit strategies helps increase transit ridership, reduce car trips, and improve overall system efficiency.

Electric vehicles (EVs) and other low-carbon fuels represent a third area of coordinated strategy. State policies and funding have pushed cities and regions to prioritize EV readiness. Fremont, for example, incorporated EV charger requirements into its zoning code and coordinated across departments to ensure new multifamily buildings are prepared for a zero-emissions future. Across the analyzed CAPs, 18 of 20 cities include EV strategies ranging from charging station networks to infrastructural readiness, while regional plans emphasize financing, public chargers, and electrification of transit fleets and goods movement. Low-carbon fuels complement these strategies, often intersecting with vehicle electrification, fleet management, and goods movement to further reduce GHG emissions.

Emerging technologies, such as autonomous vehicles (AVs), are beginning to shape transportation planning. SCAG and MTC/ABAG embed AV forecasts and data modeling into their SCSs, while SJCOG and SACOG reference AVs in long-term planning. Local CAPs in San Jose and Los Angeles also consider AVs as part of a strategy to position themselves as leaders in new transportation technologies, demonstrating how local innovation can complement regional visioning.

Goods movement is another critical focus. Regional plans in SCAG, SANDAG, and other MPOs aim to reduce GHG emissions across freight networks – from highways and railroads to last-mile delivery – through infrastructure improvements, electrification, and technology adoption. At the local level, half of the analyzed cities incorporate goods movement strategies in CAPs, optimizing energy use and addressing emissions from delivery trucks, including idling near schools, childcare facilities, or commercial hubs.

Finally, climate-friendly infrastructure often intersects with active transportation and adaptation strategies. Green streets, reflective pavements, shade vegetation, and energy-efficient lighting improve the walking and biking experience while capturing carbon and mitigating urban heat. Fourteen of 20 cities include such strategies, with cities like San Diego, Fremont, and Escondido embedding these interventions in climate adaptation plans. These efforts show that environmental co-benefits – safety, comfort, and resilience – can be achieved when transportation and infrastructure strategies are coordinated across scales.

Together, these strategies demonstrate that alignment between regional and local plans matters – but so does local autonomy. When cities have the tools, vision, and support to lead, they can push the envelope in ways that regional plans alone cannot. The result is a transportation system that is safer, cleaner, and more equitable, advancing climate goals while meeting the needs of diverse communities.

3.2 Land use: The foundation beneath it all

Land use is the backbone of any transportation strategy. If people cannot live near where they work, shop, or access daily services, it is difficult to reduce driving and achieve meaningful GHG reductions. **Table 2** summarizes the adoption of key land use strategies. Across California, regional and local plans consistently emphasize strategies like transit-oriented development (TOD), infill housing, and improving the jobs-housing balance. How cities implement these strategies, however, can vary widely.

Take San Diego. Its "City of Villages" approach doesn't simply call for density – it paints a vision of walkable, mixed-use neighborhoods anchored by transit. The city's CAP explicitly links this strategy to climate goals, guiding zoning updates, infrastructure investments, and community development. Fremont, by contrast, focuses on housing within walking distance of "basic services," while Oakland emphasizes proximity to "essential destinations of everyday life," and Sacramento plans new residential neighborhood centers. While the language and specific focus differ, all three approaches aim to shorten distances between housing and key destinations, reducing the need for car travel.

Santa Monica offers a complementary example of local innovation. Rather than focusing solely on where development occurs, the city examines how parking policies shape car use. By eliminating minimum parking requirements and adjusting rates based on demand, Santa Monica has begun to reshape how residents and developers think about driving, demonstrating that relatively small land-use adjustments can have outsized effects on transportation behavior.

Strategy	SCSs adopting (n = 5)	% of SCSs	CAPs adopting (n = 20)	% of CAPs
Transit-oriented development (TOD)	5	100%	17	85%
Infill development	5	100%	16	80%
Accessory dwelling unit (ADU) programs	2	40%	2	10%
Housing near activity centers	5	100%	14	70%
Affordable/jobs-housing balance	5	100%	12	60%
Preserve open space	5	100%	10	50%
Urban growth boundary (UGB)	1	20%	0	0%
Parking requirements	2	40%	16	80%
Urban forest	1	20%	16	80%
Port policies	1	20%	4	20%

Table 2.
Adoption of land use strategies across MPOs and cities in California.

These narratives reflect broader patterns in the data. **Table 2** shows that TOD is adopted in 17 of 20 analyzed cities, infill development in 16, housing near development centers in 14, and housing affordability and jobs-housing balance in 12. The prevalence of these strategies across SCSs and CAPs suggests a degree of vertical integration from state to municipal planning, though local implementation details differ based on city size, geography, and local priorities. Midsize and smaller cities, like Fremont and Escondido, often integrate these strategies into economic development efforts, while larger cities, such as San José, Oakland, and San Francisco, consider commuting patterns, housing affordability, and equity in linking residents to local jobs.

Preservation of open space, farmland, and critical environmental areas is another recurring theme. All five MPOs address this strategy, and roughly half of the analyzed cities incorporate it into CAPs. Approaches vary: Sacramento frames preservation as part of sustainable growth and project design, while San Francisco and Los Angeles pursue "healthy ecosystems" that integrate urban and natural environments. Oakland emphasizes equity, reflecting how land-use strategies can be tailored to community priorities.

Parking reform is emerging as a distinct climate strategy. Sixteen of 20 CAPs include measures to reduce or eliminate minimum parking requirements, often paired with innovations like shared parking, curb-space optimization, or EV-preferred parking. Regional efforts, notably in SCAG and MTC/ABAG, include legislative and planning initiatives that complement these local actions, such as dynamic parking fees designed to discourage solo driving.

Urban greening, while less common at the regional scale, is widely implemented locally, with 16 cities adopting strategies that enhance active transportation infrastructure, traffic calming, street surface cooling, and trail connectivity. Urban growth boundaries are even rarer; only MTC/ABAG at the regional level and San Jose, via its General Plan, address this approach.

Finally, land use intersects with goods movement infrastructure. Ports can be major sources of GHG emissions, yet most regional and municipal climate plans provide limited coverage. SANDAG's SCS integrates port electrification policies, and cities like Oakland and Long Beach mirror this approach in their CAPs, emphasizing drayage truck electrification or hydrogen fuel. Los Angeles focuses on both ship fuel and drayage emissions, while San Francisco and San Diego's ports, as separate planning agencies, have developed their own CAPs.

The broader takeaway is clear: while some strategies, like TOD, infill, and housing near activity centers, are widely adopted and often coordinated across scales, local governments are frequently more nimble, experimental, and ambitious. Cities can tailor approaches to local conditions, push forward innovative solutions like parking reform or urban greening, and advance equity in ways that regional plans alone cannot. In this sense, effective land-use planning forms the essential foundation beneath all transportation strategies.

3.3 Changing behavior: Demand management and public engagement

TDM refers to strategies that reduce driving, not by banning cars, but by making alternatives more attractive and accessible. These strategies often combine education, incentives, employer engagement, and infrastructure improvements to influence travel behavior. **Table 3** shows the adoption of TDM strategies across California.

Strategy	SCSs adopting (n = 5)	% of SCSs	CAPs adopting (n = 20)	% of CAPs
TDM programs	5	100%	18	90%
Other programs or incentives to lessen driving	2	40%	5	25%
Education and outreach	2	40%	12	60%

Table 3.
Adoption of TDM strategies across MPOs and cities in California.

In Roseville, for example, the city created a "Green Commute" program offering cash incentives for carpooling, biking, and taking transit. The effort went beyond financial motivation: the city launched a comprehensive public campaign to make green commuting part of its identity, using billboards, social media, and school outreach. Participation and emissions reductions were tracked annually, tying behavior-change initiatives directly to climate goals. Lodi takes a different, complementary approach by promoting telecommuting and flexible work schedules, such as four-day workweeks, to reduce congestion and vehicle trips.

These local initiatives reflect broader patterns across California. All MPOs and nearly all cities (18 of 20) incorporate TDM strategies, as shown in **Table 3**. Regional plans often include transportation system improvement policies embedded in their forecasts or visions, emphasizing technological solutions to monitor and manage traffic flow in real time across all modes. These improvements work synergistically with TDM actions, making commutes more efficient and amplifying the impact of behavioral interventions.

Education and outreach are also common elements. Twelve of the 20 analyzed cities include strategies to teach residents about sustainable travel options, ranging from bike and walk encouragement programs to alternative transportation pilots and collaborative partnerships. Medium and smaller-sized MPOs, such as SACOG and SANDAG, incorporate regional education and outreach efforts as well, with SANDAG explicitly recommending that local governments expand the policy spectrum "beyond traditional TDM strategies" to include shared streets, shared mobility pilots, micromobility incentives, and technology-based solutions.

Our analysis shows that TDM strategies are most effective when combined with supportive transit and land-use planning [7, 11]. When CAPs and SCSs coordinate TDM with infrastructure investments and strategic zoning, cities see measurable reductions in single-occupancy vehicle trips and gains in active transportation. The takeaway is clear: coordinated, locally informed behavior-change programs can multiply the benefits of transportation and land-use policies, helping cities achieve both mobility and climate goals.

3.4 Cross-cutting issues: Collaboration, community, and equity

Behind every policy is a process – and the best plans don't just look good on paper; they reflect real voices. Regional collaboration, community engagement, and attention to equity are critical cross-cutting elements that shape the effectiveness of climate action planning. **Table 4** shows the adoption of cross-cutting strategies.

Strategy	SCSs adopting (n = 5)	% of SCSs	CAPs adopting (n = 20)	% of CAPs
Regional collaboration	5	100%	18	90%
Other	0	0%	3	15%
Community involvement and outreach (CIO)	4	80%	15	75%
Equity	5	100%	7	35%

Table 4.
Adoption of cross-cutting strategies across MPOs and cities in California.

San Francisco's CAP exemplifies equity-focused engagement. The city created a separate outreach appendix documenting efforts to work with non-English-speaking communities, neighborhood groups, and advocacy coalitions. Feedback was not simply collected; it actively shaped the plan. Oakland provides another model. Its Equitable Climate Action Plan (ECAP) recentered the planning process around frontline communities – those facing intersecting racial, environmental, and socioeconomic vulnerabilities – and traced ECAP actions across all local planning documents. These examples show that equity requires intentionality, dedicated resources, and accountability.

Regional agencies often provide the infrastructure and technical support that enable local action. All five MPOs include strategies for collaboration with municipalities and state entities. SANDAG, for example, has offered climate action planning support to member jurisdictions since 2016, including GHG inventories and no-cost planning staff assistance. As a result, 19 of 20 SANDAG member agencies have developed a CAP. SANDAG also publishes the Regional Framework for Climate Action Planning (ReCAP), which provides templates, best practices, and monitoring guidance. Across California, 18 of 20 local CAPs include strategies for regional collaboration, reflecting the importance of shared frameworks and coordinated planning.

Community involvement and outreach (CIO) is equally widespread. Four SCSs include CIO explicitly, while MTC/ABAG maintains a separate participation plan. Federal law, including the FAST Act (2015), requires MPOs to engage all stakeholders in regional planning and coordinate with member jurisdictions. Consequently, all five regional agencies maintain Public Participation or Public Involvement Plans. Strategies frequently emphasize tribal participation, enhanced online engagement, and on-site outreach in disadvantaged communities to bridge the digital divide. At the municipal level, 15 of 20 cities incorporate CIO strategies in their CAPs, extending efforts to partner with community actors in various GHG reduction programs, campaigns, and initiatives.

Equity remains the area with the least alignment between regional and local plans. While all MPOs reference equity in SCSs, only 7 of 20 cities make it a core component of their CAPs. Larger cities are more likely to include equity strategies, reflecting the resources and capacity needed for meaningful engagement. Oakland's ECAP illustrates the potential: it defines frontline communities, integrates equity into all planning strategies, and explicitly connects equity goals to other local planning documents. These examples highlight that, while regional guidance can provide frameworks and technical support, local governments must intentionally design processes to ensure equitable outcomes.

Together, collaboration, community engagement, and equity form the connective tissue of climate action planning. They ensure that policies are not only technically sound but also socially grounded, reflecting the voices of those most affected while leveraging regional coordination to amplify local impact.

4. Discussion: What this all means

The findings from California's climate and transportation planning efforts reveal several important lessons about how regional and local strategies interact – and where the real impact comes from.

First, alignment matters, but it's not everything. When regional and local plans coordinate on strategies such as transit investment, active transportation, or EV readiness, the outcomes are amplified. Yet, some of the most innovative, community-driven approaches emerge from the local level, independent of strong regional mandates. Santa Monica's parking reforms, Oakland's equity-centered ECAP, and Fremont's proactive EV policies illustrate how cities can lead boldly, tailoring strategies to local needs and opportunities.

This tension between coordination and autonomy mirrors long-standing debates in governance theory. Multilevel governance emphasizes shared authority across tiers; polycentric governance highlights overlapping centers of decision-making; and subsidiarity suggests that decisions should be taken at the lowest feasible level [12, 13]. California's experience demonstrates that effective climate governance requires balancing these logics – leveraging regional frameworks for coherence while preserving local autonomy for innovation.

The policy implication is that governance frameworks must translate into actionable guidance: MPOs should standardize monitoring templates to ensure consistency across regions; state legislation should incentivize the inclusion of equity components in CAPs; and local governments should prioritize reforms, such as parking management, that are often politically difficult but highly impactful.

Second, coordination multiplies impact for certain strategies. TOD, infill housing, and bicycle infrastructure deliver outsized benefits when implemented consistently across scales. When these strategies are paired with TDM programs – like Roseville's Green Commute – they become even more effective, reducing vehicle trips while increasing equitable access to mobility options.

Third, cities need support. Planning capacity is uneven; smaller cities often lack dedicated staff, funding, or data resources to implement ambitious plans. Regional agencies can help level the playing field through technical assistance, shared templates, and guidance – as SANDAG's ReCAP framework demonstrates. At the state level, funding streams and directives could be tied to measurable outcomes, with equity requirements embedded as a condition for support.

Finally, measurement and accountability are essential. Tracking which strategies appear in which plans – and linking them to real-world outcomes – transforms good intentions into tangible results. Cities and regions that monitor participation, emissions reductions, and travel behavior, like Roseville and San Francisco, can refine programs over time, ensuring that climate and transportation goals are met efficiently and equitably. A clear policy takeaway is that MPOs and state agencies should develop standardized reporting templates and shared metrics to improve comparability, reduce administrative burdens for smaller cities, and ensure accountability across scales.

Together, these lessons underscore a simple truth: achieving sustainable, low-carbon transportation requires both top-down guidance and bottom-up innovation. Coordination accelerates change, but local creativity, capacity, and engagement ultimately determine whether strategies succeed in practice.

5. Conclusions

California's experiment in coordinated climate planning offers lessons not just for other US states but for any place working to decarbonize transportation. SB 375 created a bold framework – but the real test lies in what happens after the legislation passes.

This chapter shows that alignment between regional and local plans matters – but so does autonomy. When cities have the tools, vision, and support to lead, they can advance strategies that large regional plans cannot easily specify. At the same time, shared goals and clear frameworks ensure that local innovations feed into a coherent statewide trajectory toward reduced greenhouse gas emissions.

Reducing emissions from transportation is ultimately about people – how they move, how they live, and how governments at every level collaborate to make communities more connected, equitable, and resilient. From this integrated analysis, four key takeaways emerge:

1. **Alignment matters most for infrastructure-heavy strategies.** Investments in transit, bike and pedestrian networks, and EV infrastructure tend to be better aligned across scales because they are concrete, fundable, and highly visible. State agencies and MPOs should expand funding programs and coordination mechanisms that prioritize cross-jurisdictional infrastructure investments, ensuring that local projects contribute to regional networks and statewide emission-reduction goals.

2. **Local innovation fills policy gaps.** Cities such as Fremont, Santa Monica, and Roseville adopt detailed, implementable policies not explicitly mandated in SCSs. State and regional frameworks should explicitly create space for experimentation and diffusion of local best practices – for example, by developing clearinghouses of local innovations or offering grants that reward replicable, community-driven initiatives such as parking reform or TDM pilots.

3. **Capacity shapes adoption.** Smaller jurisdictions rely heavily on regional tools and resources – SANDAG's ReCAP program is a leading example of how MPO support can accelerate local planning. Regional agencies should standardize monitoring templates, provide shared technical resources, and bundle data services to reduce burdens on smaller cities. At the state level, funding could be tied to capacity-building, ensuring that less-resourced jurisdictions are not left behind.

4. **Equity requires intentionality.** While all regional plans mentioned equity, only a minority of CAPs embedded it as a core strategy, underscoring the need for stronger mandates, funding, and accountability. Smaller municipalities, in

particular, face challenges integrating equity, likely due to limited resources and technical capacity at the time these plans were developed. In recent years, state and regional efforts have begun to address these gaps through targeted supports. Expanding and sustaining these measures would help smaller cities integrate equity systematically into climate and transportation planning, ensuring benefits reach all communities. Moving forward, building on existing state efforts to embed equity requirements into climate and transportation mandates, MPOs should provide standardized tools such as detailed and context-sensitive equity mapping frameworks and indicators. Local governments should institutionalize equity offices or advisory bodies to ensure sustained community engagement and accountability.

Ultimately, SB 375 set the stage for the vertical integration of transportation and land use planning, but the depth of that integration depends on local capacity, political will, and sustained regional and state support. The California experience suggests that meaningful decarbonization will always be a dual project: collective in vision, local in execution.

Acknowledgments

The authors gratefully acknowledge partial support from the California State University Transportation Consortium, funded through the State of California's Road Repair and Accountability Act of 2017, as well as support from Northeastern University and the Spatial Analytics and Visualization Institute (SAVI) at San José State University. The authors also thank graduate student Branka Tatarevic for her assistance with data collection and analysis and acknowledge the use of ChatGPT to support language refinement and clarity in this chapter.

Author details

Serena E. Alexander[1*] and Ahoura Zandiatashbar[2]

1 Northeastern University, Boston, MA, United States of America

2 San José State University, San José, CA, United States of America

*Address all correspondence to: se.alexander@northeastern.edu

References

[1] Schoradt B. Sustainable communities strategies will be essential to the success of SB 375. Ecology Law Quarterly. 2009;**36**(2):611–614. Available from: https://www.jstor.org/stable/44321171

[2] Barbour E. Evaluating sustainability planning under California's senate bill 375. Transportation Research Record: Journal of the Transportation Research Board. 2016;**2568**(1):17–25. DOI: 10.3141/2568-04

[3] Frick KT, Chapple K, Mattiuzzi E, Zuk M. Collaboration and equity in regional sustainability planning in California: Challenges in implementation. California Journal of Politics and Policy. 2015;7(4):1–25. DOI: 10.5070/P2cjpp7428924

[4] Garde A, Kim C. Form-based codes for zoning reform to promote sustainable development: Insights from cities in Southern California. Journal of the American Planning Association. 2017;**83**(4):346–364. DOI: 10.1080/01944363.2017.1364974

[5] Margerum RD, Lewis R, Dobrinich S. Collaborative metropolitan governance: An evaluation of local adoption in Denver. Cities. 2019;**93**:136–144. DOI:10.1016/j.cities.2019.04.010

[6] Allred D, Chakraborty A. Do local development outcomes follow voluntary regional plans? evidence from Sacramento region's blueprint plan. Journal of the American Planning Association. 2015;**81**(2):104–120. DOI: 10.1080/01944363.2015.1067574

[7] Zandiatashbar A, Alexander S, Tatarevic B. What tackles vehicle GHG emissions in California: Regional plan adoption or local leadership? Sustainability. 2025;**17**(3):1213. DOI: 10.3390/su17031213

[8] Bartholomew K, Ewing R. Building Planner Commitment: Are Oregon's SB 1059 & California's SB 375 Models for Climate-Change Mitigation? NITC-RR-868. Portland, OR: Transportation Research and Education Center (TREC); 2017. DOI: 10.15760/trec.188

[9] Sciara GC. Regional land use and transportation sustainability: Metropolitan planning organizations (MPOs) and the art of nudging local governments. Paper presented at the Transportation Research Board 98th Annual Meeting, Washington, DC; 2019, 13–17

[10] California Air Resources Board. Sustainable communities and climate protection program [Internet]. 2025. Available from: https://ww2.arb.ca.gov/our-work/programs/sustainable-communities-climate-protection-program [Accessed: 2025-October-02]

[11] Alexander SE, Zandiatashbar A, Tatarevic B. Fragmented or aligned climate action: Assessing linkages between regional and local planning efforts to meet transportation greenhouse gas emissions reduction targets. Mineta Transportation Institute. 2022. Available from: https://transweb.sjsu.edu/research/2146-Alexander-Climate-Action-Planning-Transportation-Emissions-Senate-Bill-375 [Accessed: 2025-October-02]

[12] Hooghe L, Marks G. Types of Multi-level Governance. In Enderlein H, Wälti S, Zürn M, editors. Handbook on Multi-level Governance. Cheltenham: Edward Elgar Publishing; 2010:17–31. DOI: 10.4337/9781849809047.00007

[13] Nesting MGR. Subsidiarity, and community-based environmental governance beyond the local level. International Journal of the Commons. 2008;**2**(1):75–97. Available from: https://www.jstor.org/stable/26522991

Chapter 5

Climate-Smart Soybean Production in Nigeria: Integrating Agronomy, Risk Financing, Infrastructure and Clean Energy for Food System Resilience

Adeyinka Richard Aroyehun, Adanna Henri-Ukoha and Elizabeth Ekaette Oladejo

Abstract

Climate variability poses substantial challenges to agricultural production, particularly in developing countries, including Nigeria, where small-scale farming predominates. This study examines the influence of climate change on soybean production in Nigeria from 1961 to 2022 using autoregressive distributed lag, error correction, and vector autoregressive models. The findings indicate that climate change and land use exert positive short-run and long-run effects on soybean yields. Diagnostic tests confirmed the stability of the model, with no evidence of heteroscedasticity or autocorrelation, and normally distributed residuals. The CUSUM and CUSUM of Squares tests further validated parameter stability. Irrigation, employed as a proxy for agronomic practices, revealed limited coverage and modest infrastructure development, while transportation infrastructure remained relatively stable. Crop insurance was found to be insufficient, reflecting restricted access to climate finance. Although soybean production was not a direct contributor to clean energy supply, it indirectly supported emission stabilization. Soybean is a vital legume for food security, animal feed, and soil fertility enhancement. The study recommends a comprehensive strategy that integrates climate-resilient practices, improved agronomic techniques, sustainable water management, expanded crop insurance, and promotion of clean energy. Embedding these measures into agricultural policy will strengthen food security, resilience, and sustainable development in Nigeria.

Keywords: ARDL, climate change, environmental sustainability, food security, insurance, soybean yield

IntechOpen

1. Introduction

Ensuring food security and sustainability for impoverished farmers and consumers in Sub-Saharan Africa, who live on or below $2 per day, remains a crucial challenge, exacerbated by nutritional deficiencies [1] and climate change. In addition to the pressure faced by impoverished rural inhabitants, the emerging needs of urban Sub-Saharan African (SSA) middle-income consumers, who strive for higher dietary quality, food consumption, and higher-quality animal proteins, must also be addressed. These nutritional needs may be influenced by factors such as value, cost, source (plant or animal origin), consumer perception, and other socioeconomic or psychographic characteristics. However, the agricultural sector is responsible for meeting the demands of the rapidly rising population, which has an increased need for protein-rich diets. Small-scale farmers are the primary group, managing around 80% of farmland and producing approximately 70% of the food consumed in Africa [2].

The agriculture sector is responsible for meeting the needs of a fast-expanding population with a rising hunger for protein. Smallholder farmers constitute the foundation of African agriculture, accounting for 80% of cropland and 70% of food production [2]. African producers continue to face low economic returns and compromises between subsistence and commercial crop cultivation due to resource constraints and insufficient technology [3]. The availability and accessibility of agricultural inputs and resources impede much-needed increases in productivity and crop and farm animal choice. More effective training, easier access to funding, and, most importantly, high-quality factors of production (which comprise the quality of seeds and other inputs) are needed to eliminate the causes of the generally known yield gap – that is, the difference between the maximum attainable harvest rates and actual yields [4] – which is experienced by producers across much of Africa [5]. Many African nations have relatively low agricultural production and substantial yield disparities, which are due in part to environmental restrictions such as poor soil quality and limited water availability. Furthermore, a lack of technology, inadequate institutions and/or weak governance, land tenure, deteriorating infrastructure, and inadequate marketing or market structures force producers to undertake risk-reducing cropping decisions. The continuance of production disparity and cycles of agrarian poverty in Africa remains obvious [3].

Climatic variability refers to variations in climatic patterns over a specific time frame, which can have a considerable influence on agricultural production. In Nigeria, where the agricultural sector is a core component of the national economy and employs a large proportion of the population, understanding the impacts of climatic variability on crop output is critical for ensuring food security and sustainable development [6]. Soybean (Glycine max) has emerged as an important crop in Nigeria due to its substantial protein content and adaptability in a variety of culinary and animal feed applications. However, rising temperatures, variable precipitation patterns, and severe weather, including extreme events such as flooding and drought, pose major challenges to soybean agriculture. Climate variability affects not only crop yields but also the health of the soil, disease and pest patterns, and access to water, posing a challenge to the sustainability of food production systems. Crop management and other strategies that can mitigate the negative effects of climate change on Nigeria's key food crops are necessary for the agricultural industry to produce food sustainably. As a result, research on climatic factors and their influence on important crops such as soybeans, which is currently lacking in Nigeria, has the potential to increase food production and contribute to SDG 2's Zero Hunger. The purpose of this study is to examine how climatic variability affects soybean output in Nigeria, with a focus on the consequences for food supply sustainability. By analyzing historical climate data, we can identify soybean yield patterns and opportunities created by climate change. Finally, this study contributes to the discussion on sustainable agricultural practices by providing insights for policymakers and various stakeholders, including farmers, in their efforts to achieve food security amid changing climate factors.

Figure 1 illustrates the conceptual structure for climate-smart soybean farming. It identifies five critical components or pillars that work together to enhance resilient, sustainable soybean production in the face of climate change. Agronomy techniques, financial risk management, infrastructure development, clean energy solutions, and food system resilience are the five essential elements that will promote sustainable and climate-resilient soybean production. Each aspect makes a distinct

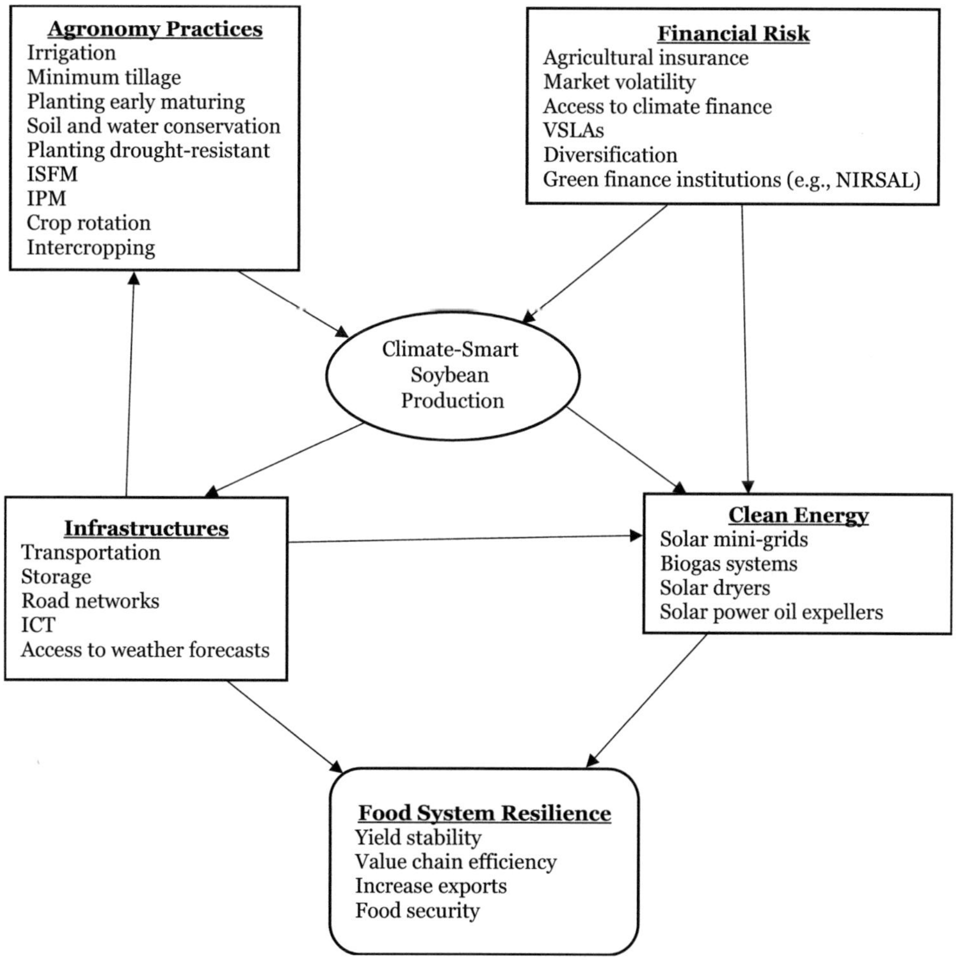

Figure 1.
Conceptual framework of the interactions between climate-smart soybean production with linkages in Nigeria. Source: authors' illustration based on data from [10–12].

contribution to climate change adaptation and mitigation. Agronomic processes are essential for climate-smart agriculture (CSA). Irrigation, limited tillage, and conservation of water and soil all contribute to mitigating drought and degradation risks [2]. Early-maturing and drought-tolerant cultivars improve adaptation to shortened harvest times [3]. Integrated soil fertility management (ISFM) and integrated pest management (IPM) increase production while lowering input costs and reducing the impact on the environment. Crop rotation and intercropping improve soil quality while reducing the prevalence of pests and the threat of crop failure [7]. CSA necessitates lowering farmers' vulnerability to financial shocks. Agricultural risk insurance and climate financing accessibility are critical tools for controlling weather and market risks. Market instability and insufficient financial access impede productivity; Village Savings and Loan Associations (VSLAs) serve as community-level alternatives. Institutions such as NIRSAL encourage green financing and reduce the risk of investments in agriculture. Diversification of revenue streams also helps to stabilize livelihoods during periods of climate instability [8]. Adequate distribution, storage, and transportation routes improve accessibility to inputs and markets, lowering postharvest losses [6]. Information and communication technologies and early warning mechanisms, such as weather predictions, facilitate informed decision-making and adaptable farming methods [9]. Thus, dependable infrastructure is a major facilitator of climate-resilient ecosystems. The use of clean energy sources may greatly reduce emissions and increase profitability in soybean production. Solar microgrids and biogas generation provide dependable, independent energy to rural areas. Solar dryers improve product quality and prevent spoilage, while solar-powered oil expellers provide environmentally friendly processing solutions. These technologies are critical for the energy shift in agriculture. The core goal of climate-smart soybean production is to enhance food system resilience, as measured by output stability, value chain efficiency, increased exports, and improved food security [7]. Hence, when soybean farming is well-integrated, it helps to diversify diets, boost rural incomes, and ensure national food security. This paradigm is for sustainable soybean farming. It emphasizes the linkages between agronomic, financial, infrastructural, energy, and systemic elements for sustainable development in the midst of changing climates.

2. Literature review

Nigeria, like many other SSA nations, is facing increasing challenges in maintaining agricultural output as a result of climate change. Soybeans, a leguminous crop recognized for its nutritional content and economic value, are identified as a critical crop for improving food security and livelihood sustainability in Nigeria [7]. CSA provides a strategy to alter soybean production systems by combining sustainable agronomic techniques, resilient financial mechanisms, infrastructure development programs, and clean energy solutions. As a result, integrating these components has the potential to establish resilient food systems in Nigeria.

2.1 Agronomic techniques in climate-smart soybean production

Agronomy is a critical component of CSA, especially in soybean production, where effective resource utilization and tolerance to climatic variations are required.

In Nigeria, CSA-compliant agronomic treatments prioritize strengthening seed systems, conserving soil and water, optimizing nutrient usage, and incorporating ecological management methods [8]. These measures increase production, strengthen resistance to climatic shocks, and ensure the long-term viability of soybean agricultural systems. The use of climate-resilient soybean cultivars is an important agronomic adaptation method. The IITA has created improved cultivars, which include TGx 1448-2E, TGx 1951-3 F, and TGx 1835-10E, that are resilient to drought, resistant to disease, and mature early [8]. These types are particularly well-suited to Nigeria's Sudan and Guinea savannah zones, where rainfall unpredictability threatens traditional soybean cultivation. Conservation agriculture, which includes no or little tillage, permanent soil cover through mulching or cultivation of cover crops, and crop rotation, improves soil structure, retains moisture, and increases organic material content [9]. These advantages improve soybean plant tolerance to drought conditions and increase total yields. ISFM involves the strategic use of organic resources (such as compost or manure), synthetic fertilizers, and biological inputs such as rhizobium inoculants. Soybeans require phosphorus and molybdenum to facilitate nodulation and the fixation of nitrogen. ISFM methods have been shown in studies to increase soybean production by more than 30% in low-input farming systems in Nigeria [13, 14].

Planting time is a major determinant in soybean farming, especially under rainfed circumstances. Integrating seasonal climatic forecasting and decision-support methods (such as rainfall onset prediction apps) could assist farmers in aligning sowing with optimum moisture conditions, reducing the risk of crop loss [15]. Agronomic calendar integration with local meteorological information via mobile technologies is an emerging technique in Nigerian CSA programs aimed at smallholder farmers [16]. Crop rotation, particularly with cereals such as maize and sorghum, aids in interrupting pest and disease sequences, increases nutrient movement, and enhances production. Soybean-maize rotation is a common practice in Nigeria's Central Region and has been shown to increase maize production in the following years because of the retained nitrogen [17]. Intercropping soybeans with mixed cassava or sorghum increases production stability, particularly under land restrictions, and can improve household food security and income. Climate change has altered the spread and intensity of pests, including soybean rust, aphids, and leaf miners. A climate-smart IPM strategy combines resistant cultivars, cultural techniques such as crop rotation, the prudent use of biopesticides like neem extracts, and biological control agents. These strategies minimize chemical consumption and improve the sustainability of ecosystems [18].

2.2 Financial risk and climate resilience in soybean production

Climate unpredictability and severe weather present significant risks to Nigerian soybean production, reducing the stability of yield and farmer income. Risk finance systems, which include agricultural insurance, microfinance, and savings plans, are critical components of CSA because they enable farmers to mitigate effects, facilitate rapid recovery, and engage in adaptive techniques [19]. Soybean farmers in Nigeria confront several risks, including irregular rainfall and drought, particularly in the Sudan and Sahel zones; floods in the south and central regions during extreme rainfall periods; and outbreaks of diseases and pests caused by climate change. These climate-induced occurrences cause production and income fluctuation, making soybean cultivation a high-risk activity for smallholders with low coping capability [20].

Index-based climate insurance is increasingly implemented as a feasible means of controlling climate risk. Instead of indemnity-based evaluations, rewards are triggered by predetermined meteorological factors, such as rainfall shortage or excess. This lowers costs associated with transactions and accelerates prompt reimbursement [21]. In Nigeria, organizations such as the Nigeria Incentive-Based Risk Sharing System for Agricultural Lending (NIRSAL) have examined crop insurance for soybeans and other commodities. NIRSAL's area production index insurance has covered several thousand small-scale farmers, increasing the ability of farmers to embrace new technology and climate-resilient techniques [22]. Nevertheless, acceptance remains poor because of limited understanding and confidence in insurance plans, affordability constraints, and a poor interface involving farmers and insurance providers.

Accessibility to climate-sensitive finance is crucial for building resilience. It allows farmers to invest in better seed types and fertilizers, implement water-saving irrigation methods, and increase the processing and storage infrastructure. However, financial availability is hampered by collateral restrictions, high interest rates, and gender inequities. Women farmers, who occupy a vital part in soybean production, may not have ownership rights or financial documents, restricting their ability to borrow [23]. Community-driven financial risk-management mechanisms, especially VSLAs and rotational savings clubs, can also offer protection against shocks. These informal structures are common in rural Nigeria and can be incorporated into formal CSA initiatives to boost resilience [24]. Furthermore, digital financial systems (such as mobile money) have created new avenues for distributing microfinance and insurance payments rapidly and securely, especially in rural areas [25]. Thus, risk finance is a critical component of climate-resilient soybean production. By expanding accessibility to insurance, financing, and financial education, Nigeria could assist smallholder farmers in adopting and scaling CSA practices. Providing accessible, inexpensive, and scalable financial solutions will be critical to developing sustainable resilience in the soybean value system.

2.3 Infrastructure for climate-smart soybean production

Infrastructure is critical to supporting CSA, especially in soybean production. Climate-smart infrastructure promotes sustainable agriculture, lowers postharvest losses, increases market access, and strengthens resilience to extreme weather events [26]. Soybean farming in Nigeria is primarily rainfed, leaving it susceptible to rainfall unpredictability and droughts. Funding in climate-resilient irrigation technologies, such as solar-powered drip irrigation, linked ridges, and gravity-supplied microirrigation, is critical for increasing water efficiency and mitigating the consequences of changes in climatic factors [27]. Supplemental irrigation throughout the blooming and pod-filling stages of soybean growth has been proven to significantly increase yields [26]. Postharvest losses in Nigeria's soybean industry are estimated to surpass 25%, owing to insufficient drying processes, storage facilities, and processing facilities [28]. These economic losses are worsened by excessive humidity, insect infestation, and aflatoxin exposure. Climate-smart postharvest options include hermetic storage bags (such as Purdue Improved Crop Storage [PICS] bags), solar-powered drying systems that rapidly reduce moisture levels, and community-centered silos for storage with both humidity and temperature management. Adoption of enhanced storage not only decreases spoilage but also stabilizes farmer revenue by allowing farmers to sell at higher prices [29].

Inadequate rural transportation networks impede soybean transit from fields to marketplaces and processing facilities, resulting in increased postharvest losses and lower farmer profitability. In accordance with the Nigeria Infrastructure Report [30], more than 60% of rural roads are in poor condition, restricting access to input and product markets. Investing in climate-resilient transport systems, like raised and weatherproof road networks, is crucial for providing year-round access to markets, particularly in areas susceptible to flooding. Improved connectivity also reduces transportation costs and allows for the timely distribution of seeds and fertilizers. Access to climatic and market data is critical to CSA. The advancement in electronic infrastructure, like mobile-based consulting services, weather forecasting apps, and e-extension systems, enhances the diffusion of weather predictions, agronomic recommendations, and decision-making on the timing of planting, pest incidents, and market dynamics [31]. Smallholders and cooperatives are unable to increase the value of soybeans due to a lack of processing equipment and adequate electricity infrastructure. In many rural regions, diesel-powered milling is inefficient and financially demanding, and grid energy is intermittent or inaccessible [32]. The use of solar-powered processing machinery, including grinders and oil extractors, is considered a viable option. These methods minimize emissions, reduce operating costs, and strengthen women's organizations involved in small- to medium-scale soybean milling [33]. Technology is a key facilitator of climate-smart soybean frameworks. Improving irrigation systems, storage, transportation, energy, and digital systems may improve resilience, minimize postharvest losses, and increase production throughout the entire soybean value chain. Bridging these infrastructural gaps, particularly in underserved rural regions, is critical to Nigeria's agricultural industry being economically viable, inclusive, and environmentally friendly.

2.4 Integration of clean energy into soybean value addition

The use of renewable energy within soybean production processes is an important technique for developing CSA in Nigeria. Since climate change continues to impact agricultural output via unpredictable precipitation patterns, warmer temperatures, and increased postharvest losses, renewable energy methods provide long-term solutions for increasing efficiency, lowering GHG emissions, reducing postharvest losses, and strengthening food supply resilience [26, 34]. At the production stage, energy is necessary for the preparation of the land, irrigation systems, and management of crops. Conventional methods that rely on energy from petroleum are not only environmentally unsustainable but also financially costly for smallholder farmers. Clean energy technology, such as solar-powered irrigation devices, seeders, and sprayer systems, has been found to reduce fuel costs, enhance off-grid power availability, and increase production [34]. In areas like the northern part of Nigeria, where rainfall uncertainty affects crop output, solar-powered drip and spray irrigation has demonstrated efficiency in reducing water stress during critical crop growth phases [27]. These devices increase water efficiency and can sustain dry-season soybean planting, extending production time. Postharvest losses in the growing of soybeans are considerably high due to insufficient drying, storage facilities, and other processing amenities, most of which are energy-intensive. The use of solar drying for moisture management, as well as sunlight-powered threshers and oil extraction devices for processing and value enhancement, has the potential to significantly minimize spoilage and improve the quality of the product [33]. Furthermore, the utilization of solar-powered cold storage systems promotes seed preservation and oil longevity, which retain soybean high-quality over prolonged storage periods, particularly in areas with inadequate energy availability. Such solutions minimize reliance on diesel generators, which are expensive and ecologically damaging [28].

Integration of clean energy throughout the soybean production process promotes local economic growth and green employment generation. Women's organizations and youth-led firms involved in the soybean process benefit from decentralized solar milling and the extraction of oil components, which increase revenue while reducing environmental impacts [35]. The replacement of diesel-based technologies with solar and bioenergy systems significantly reduces GHG emissions, aligning with Nigeria's Nationally Determined Contributions (NDCs) under the Paris Climate Agreement [36]. Switching to sustainable energy in soybean processing plants decreases air pollution, enhances workplace ergonomics, and increases national agripreneur sustainability credibility. Despite the established benefits, multiple barriers prevent clean energy inclusion in soybean value chains, including significant upfront capital expenditure, restricted access to finance, low knowledge and technical competence, and inadequate collaboration between the agricultural and energy industry sectors. Addressing these challenges requires a multistakeholder strategy combining the government, business sector, and development organizations to provide subsidies, loan accessibility, and capacity-building initiatives designed to expedite agro-energy incorporation [37]. As a result, including renewable energy throughout soybean value chains is critical to attaining sustainable, climate-resilient, and inclusive growth in agriculture in Nigeria. Clean energy, such as solar-powered generation, postharvest handling, and processing, increases productivity and decreases environmental impacts, while supporting rural development and economic empowerment. Sustainable financial investments, regulatory synergy, and collaboration among institutions are critical for expanding these advancements throughout Nigeria's soybean belt. Despite increased interest in CSA, information gaps persist regarding the economic viability of integrating CSA packages, their long-term impacts on ecological services, and successful scaling tactics. As a result, climate-smart soybean farming in Nigeria offers significant potential for increased resiliency, efficiency, and environmental sustainability. Nigeria can establish a strong soybean value chain, which contributes significantly to national food system sustainability by combining enhanced agronomic techniques, climate risk finance, climate-resilient structures, and clean energy technology. **Figure 1** depicts a conceptual framework – a comprehensive strategy integrating agronomic practices, risk finance, infrastructures, and clean energy – that promotes resilient, climate-smart soybean production, benefiting the farming community and advancing national food security targets.

2.5 Theoretical framework

The climate variability theory contends that fluctuations in climatic parameters such as temperature, rainfall patterns, and extreme weather events have a substantial impact on agricultural output, specifically soybean yield [38]. Variations in rainfall and temperature, according to this idea, can cause soil moisture deficits, pest outbreaks, and reduced yields, particularly in Nigeria, where rain-fed agriculture is prevalent [39]. The theory emphasizes the essentiality of adaptation techniques, which account for both short-term adverse weather conditions and long-term climate change trends, in improving food system resilience [40].

Food system resilience theory provides an additional perspective. This theory incorporates the notion of climate change impacts into the larger food system, which includes production, processing, distribution, and consumption components [41]. The sustainability of soybean production in Nigeria is determined by the food system's ability to withstand climate shocks while ensuring food security. This viewpoint holds that interventions such as legislative changes, increased market access, and sustainable farming practices are essential for enhancing the resilience of food systems during climate variability [42].

The agro-ecological theory focuses on the connection between ecological processes and agricultural activities. Climate variability changes the ecological dynamics of farming systems, affecting pest population dynamics, soil nutrient availability, and water supplies, all of which significantly affect soybean productivity [43]. This approach promotes agroecological concepts, including intercropping, organic soil management, and conservation agriculture as adaptive strategies to mitigate climatic shocks [44, 45].

According to Cai et al. [46], crop insurance helps farmers implement sustainable clean-energy methods by lowering the financial risks associated with innovation and climatic uncertainty. According to the Ecological Modernization Theory [47], technical innovation, such as soybean-based bioenergy, can help match environmental and economic aims. Soybeans serve as a bioenergy feedstock (e.g., biodiesel production), providing a renewable alternative to fossil fuels [48].

3. Methodology

3.1 Data gathering and sources

The data for this study were derived from Nigeria's yearly time series data, covering the period 1961–2022. The Food and Agriculture Organization (FAO) provided data on soybean output and area harvested [10], whereas the World Development Indicators (WDI) supplied data on domestic credit and carbon dioxide (CO_2) emissions [11]. Furthermore, the World Bank (Climate Knowledge Portal) provided data on temperature and precipitation [12]. To ensure robust results and to represent the estimates as elasticities, all variables used in the study were log-transformed. **Table 1** presents variables, codes, definitions, unit measurements, and data sources.

3.2 Model specification

The auto-regressive distributed lag (ARDL) bounds test model [49, 50] was applied to investigate the cointegration relationship among the variables. The ARDL bounds testing model is specified as follows:

Variables	Code	Definition	Measurement	Sources
Soybean	LSoyprod	Soybean production.	Tonnes	[10]
Area harvested	LAreaharv	Area of soybeans harvested.	Hectares	[10]
Domestic credit	LDomcred	Financial firms provide financial resources to the private sector in the form of loans. Financial institutions include monetary authorities, commercial banks, leasing businesses, moneylenders, insurance companies, pension funds, and foreign exchange firms.	% of GDP	[11]
Temperature	LTemp	Observed annual average temperature.	°C	[12]
Precipitation	LPrecip	Observed annual average precipitation.	Mm	[12]
Carbon dioxide emissions	LCO_2	Carbon dioxide emissions from liquid fuel use are mostly due to the usage of gasoline as a source of power.	Kt	[11]

Source: Authors' computation using data from [10–12].

Table 1.
Variable details.

$$\begin{aligned}\Delta \text{Lsoyprod}_t = \alpha_0 &+ \sum_{i=1}^{n} \alpha_{1i}\Delta Lsoypod_{t-i} + \sum_{i=1}^{n} \alpha_{2i}\Delta Lareaharv_{t-i} + \sum_{i=1}^{n} \alpha_{3i}\Delta Ldomcred_{t-i} \\ &+ \sum_{i=1}^{n} \alpha_{4i}\Delta Ltemp_{t-i} + \sum_{i=1}^{n} \alpha_{5i}\Delta Lprecip_{t-i} + \sum_{i=1}^{n} \alpha_{6i}\Delta LCO_{2t-i} \\ &+ \beta_0 Lsoyprod_{t-1} + \beta_1 Lareaharv_{t-1} + \beta_2 Ldomcred_{t-1} \\ &+ \beta_3 Ltemp_{t-1} + \beta_4 Lprecip_{t-1} + \beta_5 Lprecip_{t-1} \\ &+ \beta_6 Lco2_{t-1} + \mu_t \end{aligned} \tag{1}$$

where;
Δ is the first difference
Lsoyprod is the soybean production
Lareaharv is the area harvested
Ldmcred is the domestic credit
Ltemp is the temperature
Lprecip is the precipitation
Lco2 is the carbon dioxide emissions
μ is the error term.

The bounds analysis approach was conducted using the Wald test (F- and t-statistics). The null hypothesis (no cointegration) and the alternative hypothesis (cointegration) are specified as follows:

$$H_0 : \beta_1 = \beta_2 = \beta_3 = \beta_4 = \beta_5 = \beta_6 = 0$$

$$H_1 : \beta_1 \neq \beta_2 \neq \beta_3 \neq \beta_4 \neq \beta_5 \neq \beta_6 \neq 0$$

If the F-statistic value exceeds I(1), it can be inferred that the variables exhibit long-run cointegration. Furthermore, if the F-statistic value falls below I(0), cointegration is not present between the variables. Finally, if the F-statistic value is between I(0) and I(1), the conclusions are inconclusive. Similarly, if the t-statistic value is less than I(1), we conclude that the variables exhibit long-term cointegration. Furthermore, if the t-statistic value exceeds I(0), there is no cointegration between the variables. Finally, if the t-statistic value is between I(0) and I(1), the results are indeterminate. The ARDL error correction models (ECM) underlying this work are specified as follows:

$$\Delta L\text{soyprod}_t = \alpha_0 + \sum_{i=1}^{n} \alpha_{1i} \Delta Lsoypod_{t-i} + \sum_{i=1}^{n} \alpha_{2i} \Delta Lareaharv_{t-i} + \sum_{i=1}^{n} \alpha_{3i} \Delta Ldomcred_{t-i} + \sum_{i=1}^{n} \alpha_{4i} \Delta Ltemp_{t-i} + \sum_{i=1}^{n} \alpha_{5i} \Delta Lprecip_{t-i} + \sum_{i=1}^{n} \alpha_{6i} \Delta LCO_{2t-i} + \lambda ect_{t-1} + \mu_t$$

where;

λ denotes the speed of adjustment

ect is the error correction term, which ranges from - 1 to 0.

The variables' lag form was calculated using the final prediction error (FPE), Akaike Information Criterion (AIC), Schwarz Information Criterion (SC), and Hannan-Quinn Information Criterion (HQ) [49, 50]. The FPE, SC, and HQ identified lag order 1 as the optimal length for the variables (**Table 5**); therefore, lag order 1 was applied to all variables.

4. Results and discussion

4.1 Trend analysis of soybean production and area harvested

Figure 2, and **3** illustrate the trend charts of soybean production (tons), area harvested (ha), and CO_2 (kt).

Figure 2 presents the trend of soybean production and area harvested in Nigeria. Both soybean output (blue line) and the area harvested (orange line) have shown a steady increase over time. The area harvested indicates greater growth than output, demonstrating that while additional land is allotted for soybean growing, productivity has not always grown accordingly. From 1961 to 1980, soybean production and area harvested remained relatively stable, indicating that soybean agricultural operations were limited throughout this period. In the early 1980s, a significant increase, notably in the harvested area, suggested a policy shift, agricultural expansion initiative, or market force that resulted in more soybean planting. There were considerable variations from the mid-1980s to the 1990s, with both output and area harvested experiencing sharp declines in certain years (around 1987 and 1991), likely attributable to climatic shocks, policy modifications, or market constraints. From 2000 to 2021, there was a sustained and significant increase in both output and area harvested, with fluctuations in specific years, namely 2014, 2017, and 2020.

The harvested area was typically greater than the production area, indicating that production efficiency (yield per hectare) may suggest production inefficiencies. This shows that, despite increased acreage for soybean farming, yields were relatively low. In certain years, notably between 1988 and 2016–2018,

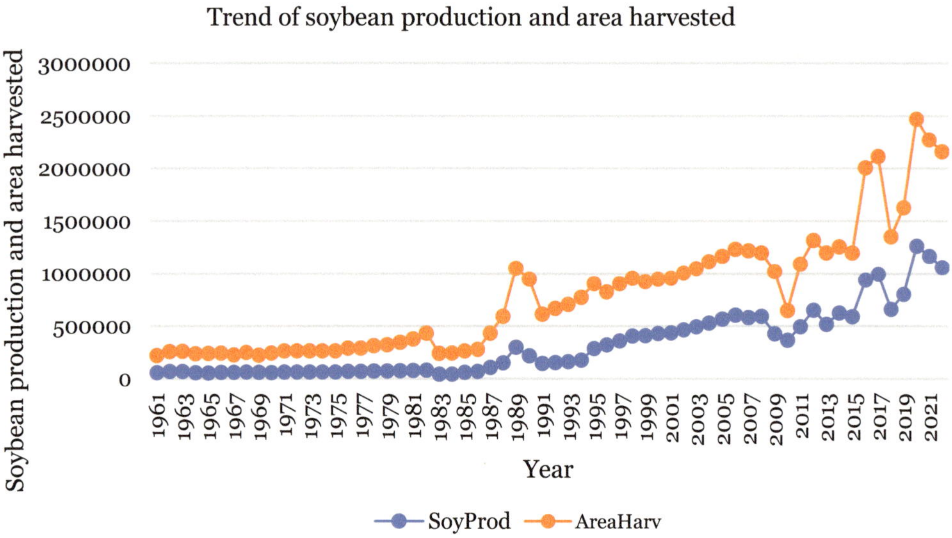

Figure 2.
Trend of soybean production and area harvested. Source: authors' computation using data from [10].

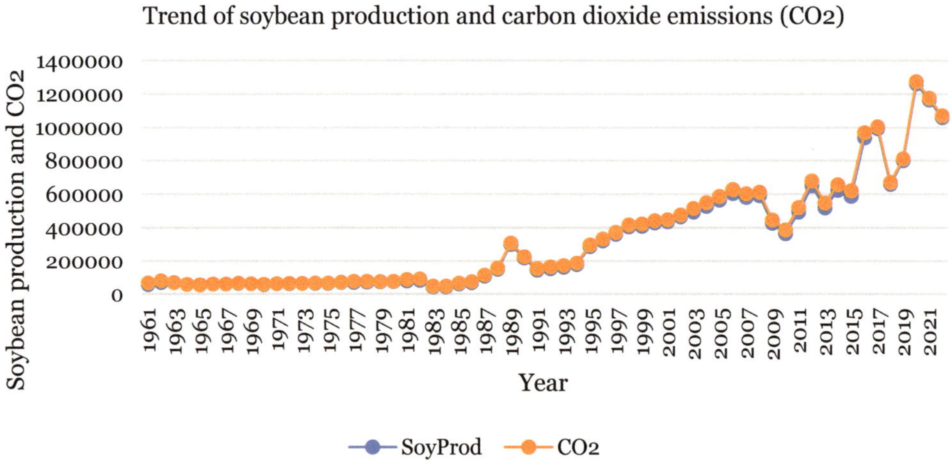

Figure 3.
Trend of soybean production and carbon dioxide emissions (CO_2). Source: authors' computation using data from [10, 11].

the gap widened significantly, indicating constraints such as pest and disease outbreaks, climate change, suboptimal agronomic practices, or input-limited accessibility. There was a significant increase in both harvested area and output, reflecting an expansion of soybean farming, which is potentially driven by demand for soy-based products (e.g., soy milk, animal feed), regulatory incentives, or global market trends, as well as indicating causal instability within soybean production systems.

4.2 Trend analysis of soybean production and CO_2 emissions

Figure 3 depicts the historical link between soybean output (SoyProd) and carbon dioxide emissions (CO_2). Both soybean output and CO_2 emissions exhibited a sustained upward trajectory. Both measures increased significantly and steadily beginning approximately in 1985. The two lines are strongly correlated, indicating a positive correlation, which means that as soybean output grows, so do CO_2 emissions. A phase of relative stability was observed from 1961 to 1980, when soybean output and CO_2 emissions were relatively low and stable, indicating limited expansion of agricultural activity or soybean-related industrial activity. The initial phase of rapid increase lasted from 1980 to 1990, with a pronounced increase in both CO_2 emissions and soybean output occurring between 1985 and 1987. This is likely attributable to a policy shift or agricultural growth (which is potentially associated with mechanization, land clearance, or fertilizer usage) and industrialization impacts from soybean processing or related infrastructure.

The period of steady and progressive increase lasted from 1990 to 2010. Soybean output and CO_2 emissions are closely aligned, implying that higher soybean production activities contribute directly to CO_2 emissions. Fertilizer usage, land clearing/deforestation, and mechanical activities (fuel combustion) are also likely contributing factors. A period of rapid growth followed by fluctuations lasted from 2010 until 2022. Soybean output increased significantly, as did CO_2 emissions. The link remains strong, with both indices peaking in comparable years (2014, 2017, and 2020). The slight declines (2020–2021) might be attributed to global factors such as market volatility, climatic shocks, or policy interventions (e.g., climate legislation and trade restrictions). As a result, the strong correlation between soybean output and CO_2 demonstrates that soybean growth contributes to CO_2 emissions. Increased farming activities (including land use change, input consumption, and mechanization) are expected to contribute to agricultural emissions. This trend reflects a carbon-intensive agriculture model in which increased output is associated with adverse environmental impacts. Increased farming activities (including land use change, input consumption, and mechanization) are anticipated to add to agricultural emissions. This trend shows a carbon-intensive agriculture paradigm in which increased output may come at the expense of environmental harm.

4.3 Summary of descriptive statistics of the variables

Table 2 summarizes the descriptive data for the factors involved in soybean production. Soyprod represents total soybean production, Areaharv represents soybean harvest area, Domcred represents domestic credit related to agricultural financial support, Temp represents an average or mean temperature affecting crop yield, Precip represents precipitation measured as average annual rainfall, and CO_2 represents the concentration of carbon dioxide, which may affect plant growth. The average Soyprod output was 324,274.1 tonnes, the average Areaharv was 466,540.3 ha, the average Domcred was 8.72% of GDP, the average temperature was 27.06°C, the average precipitation was 1174.55 mm, and the average CO_2 concentration was 10,579.05 ppm. Soyprod's standard deviation (313,578.7 tonnes) indicates substantial variability in production levels.

Soyprod (1.1719), Areaharv (0.8582), Domcred (1.1549), and CO_2 (0.8831) were positively skewed, indicating that the tail on the right side is longer or fatter than the

Variables	Soyprod	Areaharv	Domcred	Temp	Precip	CO_2
Mean	324,274.10	466,540.30	8.72	27.06419	1174.545	10,579.05
Median	170,500.00	507,250.00	7.84	27.10000	1184.070	10,141.09
Maximum	1,262,280.00	1,207,740.00	19.63	27.86000	1356.450	32,702.31
Minimum	42,000.00	160,000.00	3.70	26.27000	888.0600	58.67200
Std. Dev.	313,578.70	276,323.70	3.34	0.395746	90.56450	9053.086
Skewness	1.17	0.86	1.15	-0.170204	-0.443241	0.883119
Kurtosis	3.70	3.19	4.15	2.306063	3.417994	3.063574
Jarque-Bera	15.48	7.70	17.20	1.543352	2.481468	8.069401
Probability	0.0004	0.0213	0.0002	0.462238	0.289172	0.017691
Sum	20,104,992	28,925,497	540.49	1677.980	72,821.76	655,900.8
Sum Sq. Dev.	6.00E + 12	4.66E + 12	680.9227	9.553510	500,317.7	5.00E + 09
Observations	62	62	62	62	62	62

Source: Authors' computation using data from [10–12].

Table 2.
Descriptive statistics.

left, while temperature (-0.1702) and precipitation (-0.4432) were negatively skewed to a moderate degree. The kurtosis, which assesses the tailedness of a distribution, indicates that Soyprod (3.7049) and Domcred (4.1495) were leptokurtic. Areaharv (3.1871) and precipitation (3.4180) were mildly leptokurtic. CO_2 (3.0636) was mildly platykurtic, whereas Temp (2.3061) was platykurtic. Higher kurtosis (>3) suggests a distribution with thicker tails and a sharper peak than a normal distribution.

Jarque-Bera (JB) determines if sample data exhibits the skewness and kurtosis of a normal distribution. A higher statistic indicates a greater deviation from normality. The Soyprod JB statistic of 15.47705, with a p-value of 0.000436, shows a substantial departure from normality, necessitating a test of stationarity. The Areaharv JB statistic of 7.7004, with a p-value of 0.021275, shows that the data deviates considerably from a normal distribution, necessitating a stationarity test. The Domcred JB statistic of 17.19754, with a p-value of 0.000184, indicates that the data deviates considerably from normalcy, necessitating a stationarity test. The Temp JB statistic of 1.5434, with a p-value of 0.462238, does not deviate substantially from normality, indicating that the data is normally distributed since the p-values exceed the 0.05 significance level. The Precip JB statistic of 2.4815, with a p-value of 0.289172, does not deviate substantially from normality, indicating that the data is normally distributed since the p-values exceed the 0.05 significance level. The CO_2 JB statistic of 8.0694, with a p-value of 0.017691, departs significantly from normality, necessitating a test of stationarity.

Soyprod and Areaharv exhibit strong skewness and kurtosis, indicating a positively skewed distribution, with high values exerting an upward influence on the mean. Domcred likewise shows positive skewness and significant kurtosis, suggesting a concentration of lower values. Temperature and precipitation are both reasonably regularly distributed, with temperature exhibiting a slight negative skew. CO_2 has a positive skew and substantial kurtosis, indicating that extreme observations exert influence on the mean. As a result, a stationarity test employing both the Augmented Dickey-Fuller (ADF) and Phillips-Perron (PP) tests was utilized to assess whether a time series variable is stationary or has a unit root [51].

4.4 Unit root analysis

To examine the variables' stationarity properties, this study employed the ADF [52] and PP [50] unit root tests.

Table 3 presents the results of unit root tests for the variables using two alternative approaches, namely the ADF test and the PP test. The PP test is similar to the ADF but varies in how it handles serial correlation and heteroskedasticity. A stationary time series maintains a consistent mean, variance, and autocorrelation across time. It is critical in this modeling because nonstationary series may lead to spurious regression results.

The level ADF test of LSoyprod (log of soybean production) statistic was - 0.4168 with a p-value of 0.8992, suggesting nonstationarity (the series has a unit root). The first difference ADF test of the LSoyprod statistic yielded - 6.9546, with a p-value of 0.0000, indicating stationarity after differencing. The PP test yielded results similar to the ADF test, being nonstationary at the level but stationary after initial differencing. The ADF test of LAreaharv (log of harvested area) was nonstationary at the level (p = 0.6058) but stationary at the first difference (p = 0.000). The PP test yielded the same results: nonstationary at the level, stationary after differencing. The ADF test of LDomcred (log of domestic credit) was approximately significant at the 10% level (p = 0.0780), indicating weak evidence of stationarity, and stationary at the first difference. The PP test yielded results consistent with the ADF. Both the ADF and PP tests of LTemp (log of temperature) demonstrate stationarity at the level and first difference (p-values <0.05). Both the ADF and PP tests of LPrecip (log of precipitation) indicate stationarity at the first difference (p-values = 0.0000),

Variables	Unit root test for ADF		Unit root test for PP	
	Level	First difference	Level	First difference
LSoyprod	-0.4168 [0.8992]	-6.9546*** [0.0000]	-0.2285 [0.9285]	-7.0762*** [0.0000]
LAreaharv	-1.3394 [0.6058]	-8.7071*** [0.0000]	-0.8822 [0.7874]	-11.4320*** [0.0000]
LDomcred	-2.7107* [0.0780]	-8.0399*** [0.0000]	-2.7198* [0.0765]	-10.7916*** [0.0000]
LTemp	-3.7768*** [0.0051]	-11.7394*** [0.0000]	-3.7411*** [0.0057]	-15.9523*** [0.0000]
LPrecip	-5.7822*** [0.0000]	-12.7775*** [0.0000]	-5.8410*** [0.0000]	-21.3036*** [0.0001]
LCO_2	-1.8811 [0.3389]	-12.8370*** [0.0000]	-18,026 [0.3759]	-8.0451*** [0.0000]

*ADF stands for Augmented Dickey-Fuller; PP stands for Phillips-Perron; significant levels are indicated by ***, **, and *, at the 1%, 5%, and 10% levels, respectively.*
Source: Authors' computation using data from [10–12].

Table 3.
Stationarity of unit root test.

indicating that this variable is stationary in levels. Both ADF and PP tests on LCO_2 (log of CO_2 levels) indicate nonstationary behavior at the level but become stationary after initial differencing.

4.5 Correlation analysis

Spearman's rho is a nonparametric measure of rank correlation that determines the strength of the monotonic relationship between two variables. **Table 4** presents Spearman's rho correlation as well as the variance inflation factor (VIF), which quantifies the multicollinearity of independent variables. **Table 4** indicates that lsoyprod and lareaharv (0.930) exhibit a strong and statistically significant positive correlation with soybean output and harvested area. This indicates that an increase in harvested area is associated with an increase in soybean output. lsoyprod and ldomcred (0.717) indicate a moderately strong and statistically significant positive correlation between soybean output and domestic credit. Higher levels of domestic credit are associated with higher soybean production. lsoyprod and ltemp (0.697) show a moderate positive correlation between soybean production and temperature. lsoyprod and lprecip (0.136) show a weak and statistically insignificant positive correlation between soybean production and precipitation. lsoyprod and lCO_2 (0.841) exhibit a strong positive correlation between soybean output and CO_2 levels, suggesting that higher CO_2 levels are positively associated with soybean production.

In standard econometric practice, a VIF greater than 5 implies severe multicollinearity, while a VIF larger than 10 is considered problematic. **Table 3** shows that lareaharv (2.863) indicates moderate multicollinearity, ldomcred (1.822) indicates low multicollinearity, ltemp (1.968) indicates low multicollinearity, lprecip (1.165) indicates low multicollinearity, and lCO_2 (2.501) indicates moderate multicollinearity. The VIF scores indicate that multicollinearity does not pose a serious issue in the data.

4.6 Model specification using the ARDL framework

Table 5 presents the vector autoregression (VAR) lag order selection criteria, which facilitate the determination of the appropriate number of lags to include in a VAR model. Several criteria are reported to guide this selection. "Lag" denotes the

Variables	Lsoyprod	Lareaharv	ldomcred	ltemp	lprecip	lCO_2	VIF
lsoyprod	1						
lareaharv	0.930**	1					2.863
ldomcred	0.717**	0.587**	1				1.822
Ltemp	0.697**	0.635**	0.580**	1			1.968
Lprecip	0.136	-0.023	0.170	-0.058	1		1.165
lCO_2	0.841**	0.766**	0.627**	0.683**	-0.035	1	2.501

*****means correlation is significant at the 0.01 level (2-tailed).*
Source: Authors' computation using data from [10–12].

Table 4.
Spearman's rho correlation.

Lag	LogL	LR	FPE	AIC	SC	HQ
0	106.753	NA	1.17e-09	-3.535	-3.320	-3.452
1	305.472	348.631	3.92e-12*	-9.245	-7.739*	-8,660*
2	328.457	35.485	6.45e-012	-8.788	-5.992	-7.701
3	367.612	52.206	6.45e-12	-8.899	-4.813	-7.311
4	414.821	53.007*	5.44e-12	-9.292*	-3.916	-7.203
5	445.265	27.773	9.93e-12	-9.097	-2.430	-6.506

** indicates lag order selected by the criterion, LogL: Log-likelihood value, LR: Likelihood ratio (each test at the 5% level), FPE: Final prediction error, AIC: Akaike information criterion, SC: Schwarz information criterion, HQ: Hannan-Quinn information criterion.*
Source: Authors' computation using data from [10–12].

Table 5.
VAR lag-order selection criteria.

number of lagged terms included in the model. The likelihood ratio (LR) statistic sequentially compares the model fit across different lag lengths; a statistically significant LR value indicates the inclusion of an additional lag. The FPE evaluates the predictive accuracy of the model, while the AIC balances model fit against complexity, with lower values indicating superior model specification. The SC, also known as the Bayesian information criterion (BIC), applies a stricter penalty for model complexity. The HQ provides an intermediate penalty structure between the AIC and SC. Lag 0 serves as the baseline model without lagged terms. According to multiple criteria (FPE, SC, and HQ), Lag 1 constitutes the optimal lag order, as it yields the lowest values among the alternatives. Although the LR test favors lag 4, the convergence of FPE, AIC, SC, and HQ on lag 1 supports its selection as the most appropriate order for this study.

4.7 ARDL bounds test for cointegration

Table 6 presents the results of the bounds test, which is frequently utilized in the ARDL model to evaluate the existence of long-run relationships (cointegration) among variables. The F-statistic and t-statistic are evaluated against critical values at different significance levels to test the null hypothesis of no cointegration (long-run association). The F-statistic measures the joint significance of the lagged level variables, while the t-statistic evaluates the statistical significance of the error correction component in ECM. For each significance level, two bounds are specified: the lower bound (I(0)) assumes the absence of a long-run relationship, whereas the upper bound (I(1)) assumes its presence. At the 5% significance level, if the

Bounds test	Values	10% (0)	10% (1)	5% (0)	5% (1)	2.5% (0)	2.5% (1)	1% (0)	1% (1)
F-statistic	8.79	2.45	3.52	2.86	4.01	3.25	4.49	3.74	5.06
t-statistic	-5.67	-2.57	-3.57	-2.86	-3.99	-3.13	-4.26	-3.43	-4.60

Source: Authors' computation using data from [10–12].

Table 6.
Bounds test results.

F-statistic exceeds the upper bound, the null hypothesis is rejected, indicating a statistically significant long-run relationship. Similarly, if the t-statistic falls below the lower bound, it also confirms the presence of a long-run association. In this study, the F-statistic (8.79) is significantly higher than the upper bounds at all significance levels (10%, 5%, 2.5%, and 1%), providing statistically significant evidence of a long-run relationship. Moreover, the t-statistic (–5.67) is lower than the corresponding critical values across all significance levels, further confirming the existence of a long-run association.

4.8 Dynamic ARDL estimation results

Table 7 presents the dynamic ARDL model results. The ARDL model was applied to estimate the relationship between variables in both the short and long runs, particularly when the variables are integrated at mixed levels (I(0) or I(1), but not I (2)). **Figure 4** presents the ARDL model with the lowest AIC value, which corresponds to the most parsimonious model specification. The dynamic ARDL

Variable	Coefficient	Std. Error	t-statistic	Prob.
Lsoyprod	0.854***	0.069	12.322	0.000
Short-run results				
Lareaharv	0.749***	0.089	8.456	0.000
Ldomcred	-0.048	0.075	-0.644	0.522
Ltemp	1.188	1.963	0.605	0.548
Lprecip	0.818***	0.244	3.358	0.002
lco_2	0.034*	0.018	1.871	0.067
Long-run results				
Δlareaharv	0.923**	0.416	2.216	0.031
Δldomcred	-0.331	0.598	-0.552	0.583
Δltemp	26.256*	14.229	1.845	0.071
Δlprecip	5.598**	2.820	1.985	0.052
Δlco_2	0.229	0.175	1.308	0.197
Constant	-18.563**	8.125	-2.285	0.026
ECT(-1)	-0.146***	0.032	-4.528	0.000
R-squared	0.676	Mean dependent var		0.048
Adjusted R-squared	0.665	S.D. dependent var		0.222
S.E. of regression	0.128	Akaike info criterion		-1.221
Sum squared resid	0.955	Schwarz criterion		-1.117
Log likelihood	40.239	Hannan–Quinn criterion.		-1.180
F-statistic	457.565	Durbin-Watson stat		2.202
Prob(F-statistic)	0.000			

Source: Authors' computation using data from [10–12].

Table 7.
Dynamic ARDL results.

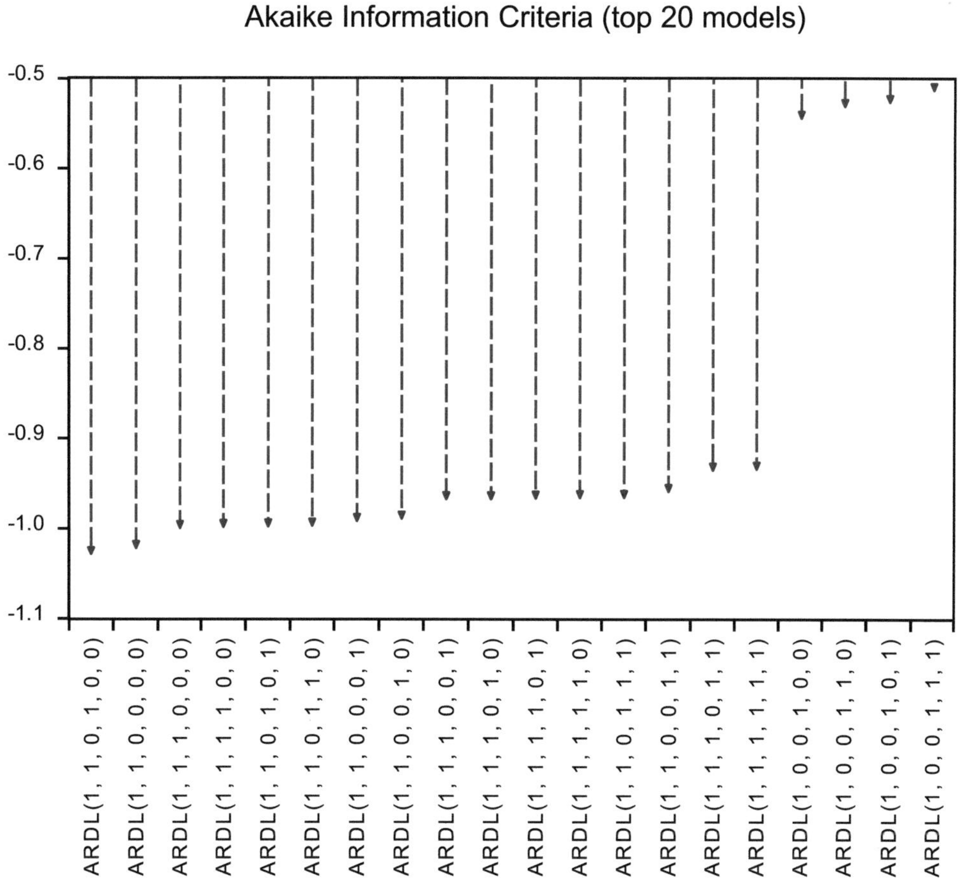

Figure 4.
Model selection. Source: authors' computations using data from [10–12].

model results for soybean report the AIC values for the top 20 models in the ARDL model selection process. The AIC is a criterion used to evaluate the goodness of fit of competing statistical models, with lower values (1, 1, 0, 1, 0, 0) corresponding to improved model fit while balancing model complexity and fit.

In the short run, the model evaluates the effects of independent variables on soybean output (lsoyprod). The log of harvested area (lareaharv) has a positive (0.749) and statistically significant effect on soybean output in the short run, such that a 1% increase in harvested area results in a 0.749% increase in soybean production. The log of domestic credit (ldomcred) is negative (-0.048) but statistically insignificant with respect to soybean output in the short run. The log of temperature (ltemp) exhibits a positive coefficient (1.188), although it is statistically insignificant for soybean yield. The log of precipitation (lprecip) has a positive (0.818) and statistically significant effect on soybean output in the short run, suggesting that increased precipitation is positively associated with soybean yield. The log of CO_2 (lco_2) is positive (0.034) and approaches marginal significance, though it is not statistically significant. These findings align with prior studies [53, 54], which reported that a warmer environment can contribute to maintaining higher soybean yields. The results are also consistent with [55–58], who reported that carbon dioxide emissions exert a significant positive effect on crop output. Conversely, other studies [59–61] found that carbon dioxide emissions negatively affect agricultural productivity.

In the long run, the model represents the relationship between the variables and soybean output after accounting for time adjustments. According to Δlareaharv (0.923), a 1% increase in harvested area is associated with a 0.923% increase in soybean yield, and this effect is statistically significant at the 5% level. The coefficient for Δldomcred (-0.331) was negative but statistically insignificant, indicating no long-run relationship between domestic credit and soybean output. The coefficient for Δltemp (26.256) indicates that temperature exerts a significant positive effect on soybean output at the 10% level, suggesting that rising temperatures may contribute to higher soybean yields in the long run. Similarly, Δlprecip (5.598) exerts a positive and statistically significant effect on soybean yield in the long run. These results are consistent with prior findings [50, 62, 63]. The coefficient for ΔCO_2 (0.229) indicates a positive but statistically insignificant effect of CO_2 levels on soybean output in the long run. The ordinary least squares (OLS) results in **Table 10** further confirmed the long-run effects of area harvested (lareaharv), domestic credit (ldomcred), temperature (ltemp), and precipitation (lprecip) on soybean output.

The error correction term (ECT) (-0.146, $p = 0.000$) indicates the speed at which the model returns to equilibrium following a short-run shock. The coefficient is negative (-0.146), highly significant ($p = 0.000$), and less than one. This indicates that 14.6% of the deviation from the long-run equilibrium is corrected each period, suggesting that the system gradually adjusts toward the long-run relationship. Model fit and diagnostic statistics show that the R-squared (0.676) explains approximately 67.6% of the variation in soybean output. The adjusted R-squared (0.665) indicates that, after adjusting for the number of predictors, 66.5% of the variation is explained. The regression standard error (0.128) measures the extent to which the predicted values deviate from the actual values. The F-statistic (457.565) tests the overall significance of the model, and its magnitude, along with the associated p-value (0.000), demonstrates that the model is statistically significant. The Durbin–Watson statistic (2.202) assesses the presence of autocorrelation in the residuals. A value close to 2 indicates the absence of autocorrelation, suggesting that the model is well-specified.

In summary, in the short run, harvested area (lareaharv) and precipitation (lprecip) exerted significant positive effects on soybean output, whereas domestic credit and temperature exhibited negligible influence. In the long run, harvested area, temperature, and precipitation exerted substantial positive effects, while domestic credit and CO_2 showed no statistically significant effect. Furthermore, the ECT indicates an intermediate speed of adjustment toward the long-run equilibrium following short-term shocks. The model exhibited a satisfactory fit and was statistically significant, with no evidence of substantial autocorrelation in the residuals. Therefore, the ARDL model demonstrates that climatic factors (precipitation and temperature) and land use (harvested area) were significant determinants of soybean output in Nigeria over time. The ECT representation equation specifying the long-term relationship between soybean production and its determinants is presented as follows:

$$ECT = LSOYPROD - (0.923{*}LAREAHARV - 0.331{*}LDOMCRED + 26.256{*}LTEMP + 5.598{*}LPRECIP + 0.229{*}LCO2 - 126.962)$$

4.9 Diagnostic tests

Diagnostic tests are statistical procedures used to evaluate the validity of regression models. They are employed to verify that the underlying assumptions of the model are satisfied, which is essential for drawing reliable inferences. **Table 8** reports the Breusch-Pagan-Godfrey test for heteroscedasticity, which examines whether the error variance is constant. The null hypothesis of homoscedasticity was not rejected since the p-value (0.120) exceeds the conventional significance level of 0.05, indicating that the error variance was constant across observations, as required for valid regression analysis. The Breusch-Godfrey LM test for autocorrelation was also conducted to assess whether regression residuals were serially correlated. The null hypothesis of no autocorrelation was not rejected, as the p-value (0.306) was greater than 0.05, indicating that the residuals were independent across time. The Ramsey RESET test was employed to detect specification errors. The null hypothesis that the model was correctly specified was not rejected, with a p-value of 0.995, suggesting no evidence of omitted variables or functional misspecification. Finally, the JB test was applied to assess the normality of residuals. The null hypothesis of normally distributed residuals was not rejected since the p-value (0.933) was above 0.05, confirming that the residuals did not deviate significantly from normality. Taken together, the results provide no evidence of heteroscedasticity, autocorrelation, or model misspecification, and confirm that the residuals were normally distributed. Overall, the diagnostic results indicate that the regression model was well-specified and met the core assumptions required for valid statistical inference.

Figures 5 and **6** present the results of the CUSUM (cumulative sum) and CUSUM of Squares tests for parameter stability. **Figure 5** illustrates the CUSUM test, which evaluates the stability of model parameters over time – a crucial requirement for the reliability of regression and time-series models in econometric analysis. The blue line represents the cumulative sum of recursive residuals, which increases or decreases in response to parameter estimation changes as additional data points are incorporated. The red dashed lines denote the 5% significance level confidence bounds, within which the parameters are expected to remain if stability holds. Since the CUSUM line remains within these bounds, the results indicate that the model parameters did not exhibit significant instability and were stable at the 5% significance level. Similarly, **Figure 6** shows the CUSUM of Squares test, which assesses parameter stability in terms of variance. The CUSUM of Squares line also remains within the 5% confidence bounds, confirming that no evidence of parameter instability was detected and that the model was stable throughout the study period.

Test	Value (Chi-square)	Probability
Breusch-Pagan-Godfrey test	1.703	0.120
Breusch-Godfrey LM test	1.214	0.306
Ramsey RESET test	0.006	0.995
Jarque-Bera test	0.139	0.933

Source: Authors' computation using data from [10–12].

Table 8.
Diagnostic test.

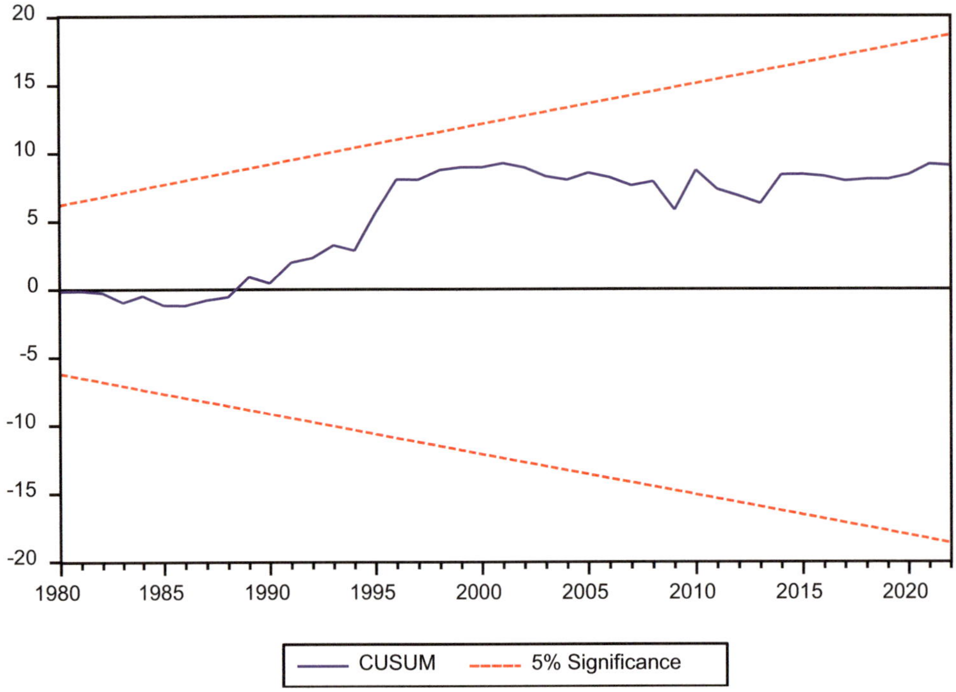

Figure 5.
CUSUM (cumulative sum) test for stability. Source: authors' computation using data from [10–12].

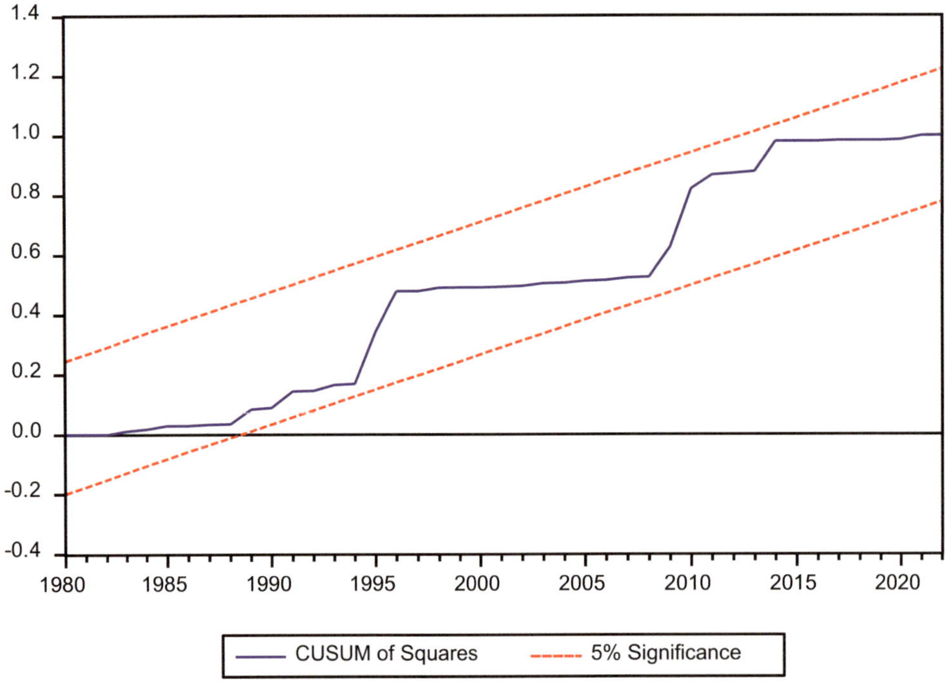

Figure 6.
CUSUM of squares (cumulative sum of squares) test for parameter stability. Source: authors' computation using data from [10–12].

4.10 Granger causality test results

Table 9 presents the results of the paired Granger causality tests at lag 2. These tests evaluate whether lagged values of one time series provide predictive information for another. The null hypothesis of no Granger causality was assessed using the F-statistic, with associated p-values reported. A larger F-statistic

S/N	Null Hypothesis (Ho)	F-Stat	Prob.	Remarks
1	LAREAHARV does not Granger-cause LSOYPROD.	0.665	0.518	Accepted
	LSOYPROD does not Granger-cause LAREAHARV.	5.453	0.007	Rejected
2	LDOMCRED does not Granger-cause LSOYPROD.	0.325	0.724	Accepted
	LSOYPROD does not Granger-cause LDOMCRED.	4.190	0.020	Rejected
3	LTEMP does not Granger-cause LSOYPROD.	3.066	0.055	Accepted
	LSOYPROD does not Granger-cause LTEMP.	6.988	0.002	Rejected
4	LPRECIP does not Granger-cause LSOYPROD.	1.576	0.216	Accepted
	LSOYPROD does not Granger-cause LPRECIP.	0.681	0.510	Accepted
5	LCO_2 does not Granger-cause LSOYPROD.	1.548	0.222	Accepted
	LSOYPROD does not Granger-cause LCO_2	2.624	0.082	Accepted
6	LDOMCRED does not Granger-cause LAREAHARV.	1.145	0.326	Accepted
	LAREAHARV does not Granger-cause LDOMCRED.	2.187	0.122	Accepted
7	LTEMP does not Granger-cause LAREAHARV.	5.038	0.010	Rejected
	LAREAHARV does not Granger-cause LTEMP.	4.077	0.022	Rejected
8	LPRECIP does not Granger-cause LAREAHARV.	0.534	0.589	Accepted
	LAREAHARV does not Granger-cause LPRECIP.	0.246	0.783	Accepted
9	LCO_2 does not Granger-cause LAREAHARV.	2.029	0.141	Accepted
	LAREAHARV does not Granger-cause LCO_2	4.002	0.024	Rejected
10	LTEMP does not Granger-cause LDOMCRED.	1.736	0.186	Accepted
	LDOMCRED does not Granger-cause LTEMP.	5.655	0.006	Rejected
11	LPRECIP does not Granger-cause LDOMCRED.	0.014	0.986	Accepted
	LDOMCRED does not Granger-cause LPRECIP.	0.215	0.8073	Accepted
12	LCO_2 does not Granger-cause LDOMCRED.	2.407	0.0995	Accepted
	LDOMCRED does not Granger-cause LCO_2	3.889	0.0263	Rejected
13	LPRECIP does not Granger-cause LTEMP.	0.354	0.7036	Accepted
	LTEMP does not Granger-cause LPRECIP.	0.552	0.5788	Accepted
14	LCO_2 does not Granger-cause LTEMP.	6.957	0.0020	Rejected
	LTEMP does not Granger-cause LCO_2	3.376	0.0414	Rejected
15	LCO_2 does not Granger-cause LPRECIP.	0.117	0.8895	Accepted
	LPRECIP does not Granger-cause LCO_2	2.391	0.1010	Accepted

****, **, and * mean significant at the 1%, 5%, and 10% levels, respectively; dynamic ARDL (1, 1, 0, 1, 0, 0) of soybean.*
Source: Authors' computation using data from [10–12].

Table 9.
Pairwise Granger causality tests at lag 2.

Variable	Coefficient	Std. Error	t-Statistic	Prob.
Lareaharv	1.278***	0.105	12.145	0.000
Ldomcred	0.333**	0.144	2.323	0.024
Ltemp	12.269***	3.626	3.383	0.001
Lprecip	1.516***	0.516	2.936	0.005
lco_2	0.035	0.036	0.961	0.341
Constant	-56.455***	12.565	-4.493	0.000
R-squared	0.930	Mean dependent var		12.172
Adjusted R-squared	0.923	S.D. dependent var		1.069
S.E. of regression	0.296	Akaike info criterion		0.492
Sum squared resid	4.895	Schwarz criterion		0.698
Log likelihood	-9.266	Hannan–Quinn criterion.		0.573
F-statistic	148.232	Prob(F-statistic)		0.000

Source: Authors' computation using data from [10–12].

Table 10.
Linear regression for the impact of climate change on soybean production.

corresponds to a greater likelihood of rejecting the null hypothesis at the specified significance level. The probability (Prob.) values indicate the statistical significance of each relationship. For instance, the null hypothesis that LAREAHARV (log of harvested area) does not Granger-cause LSOYPROD (log of soybean production) could not be rejected (p = 0.518). Conversely, the null hypothesis that LSOYPROD does not Granger-cause LAREAHARV was rejected (p = 0.007), indicating a unidirectional predictive relationship from soybean production to harvested area. Similarly, LDOMCRED (domestic credit) did not Granger-cause LSOYPROD (p = 0.724), while LSOYPROD Granger-caused LDOMCRED (p = 0.020). For LTEMP (temperature), the null hypothesis of no causality from temperature to soybean production could not be rejected at the 5% level (p = 0.055), whereas the reverse relationship was significant (p = 0.002). No significant Granger causality was detected between LPRECIP (precipitation) and soybean production (p = 0.216 and 0.510). Likewise, no significant causal relationship was observed between LCO_2 (carbon dioxide) and LSOYPROD (p = 0.222 and 0.082). Finally, the relationship between LDOMCRED and LAREAHARV was also insignificant (p = 0.326 and 0.122).

Temperature was found to Granger-cause harvested area (p = 0.010), while harvested area also Granger-caused temperature (p = 0.022), indicating bidirectional predictive causality. No evidence of Granger causality was detected between precipitation (LPRECIP) and harvested area (LAREAHARV), with p-values of 0.589 and 0.783, respectively. Carbon dioxide (LCO_2) did not Granger-cause harvested area (p = 0.141), although harvested area Granger-caused carbon dioxide levels (p = 0.024). Temperature (LTEMP) did not Granger-cause domestic credit (LDOMCRED) (p = 0.186), whereas domestic credit Granger-caused temperature (p = 0.006). The null hypotheses of no causality between precipitation and domestic credit could not be rejected (p = 0.986 and 0.807, respectively). Similarly, LCO_2 did not Granger-cause domestic credit (p = 0.0995), while the reverse relationship was significant ($p = 0.0263$). The null hypotheses of no causality between precipitation and temperature could not be rejected ($p = 0.7036$ and 0.5788, respectively). In contrast, both null hypotheses for LCO_2 and LTEMP were rejected ($p = 0.0020$ and 0.0414, respectively), providing evidence of bidirectional Granger causality. Finally, no evidence of Granger causality was detected between LCO_2 and precipitation ($p = 0.8895$ and 0.1010).

In summary, soybean production (LSOYPROD) was found to Granger-cause several variables, including harvested area, temperature, and domestic credit. Harvested area (LAREAHARV) exhibited Granger causality with temperature and CO_2. Temperature (LTEMP) Granger-caused both harvested area and CO_2. Domestic credit (LDOMCRED) Granger-caused temperature and CO_2, while no statistically significant predictive relationship was detected with the other variables.

4.11 Soybean production and agricultural insurance coverage

Subsidized crop policy is defined as crop insurance protection for food crops, in which farmers pay 50% of the premium, and the remaining 50% is covered by the Federal and State governments of the geographical location of the farm, in a 3:1 ratio, respectively. Perils covered include loss or damage caused by fire, lightning, explosion, aircraft damage, windstorm, flood, and drought. **Figure** 7 illustrates the relationship between soybean output (SoyProd) and insurance support (ISU) from 1991 to 2021. Soybean output increased steadily from 1991 to 2007, reaching around 600,000 tonnes. Between 2008 and 2011, a marked decline was observed, potentially attributable to economic, meteorological, or policy changes. An increasing trend in soybean output reemerged in 2012 and peaked in 2019 at more than 1.2 million tons. Following 2019, a modest reduction in production occurred,

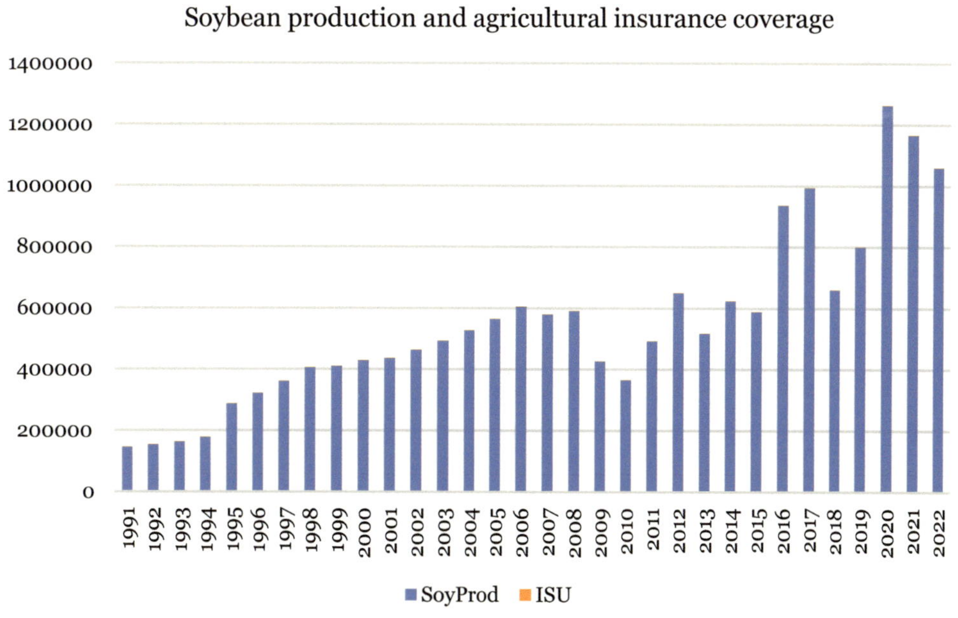

Figure 7.
Soybean production and agricultural insurance coverage. Source: authors' computation using data from [10, 11].

with production reaching roughly 1 million tons in 2021. Crop insurance values were negligible in previous years, indicating low insurance coverage. Modest increases were observed in the 2010s, although the size remains minor when compared to soybean output. This shows that ISU remains limited and insufficiently adopted, even in years of strong soybean yields. Following [64–66], the absence of awareness, insufficient risk evaluation systems, educational level, income, farm size, and limited government subsidies all contribute to the low adoption of agricultural insurance in Nigeria. Additionally, National Agricultural Insurance Commission (NAIC) data [67] show that only a negligible share of crop producers in Nigeria, fewer than 5%, are registered in official crop insurance plans. This is supported by research such as [68], which found that farmers in southern Nigeria perceive crop insurance as inaccessible or unreliable, resulting in underutilization despite increased agricultural hazards. Soybean output has increased substantially over the last three decades, but insurance coverage has not kept up. This disparity between output and insurance coverage indicates that farmers possess limited risk protection, are vulnerable to climatic shocks or market volatility, and there may be a policy gap or insufficient farmer awareness and access to insurance schemes.

As a result, Nigeria could leverage the European Central Bank's (ECB) [69] positioned institutional bond assets offered through the corporate sector purchase program to issuers that demonstrate greater climate performance. This comprises businesses engaged in reuse and recycling, environmentally sustainable design, energy conservation, and resource recovery, all of which are consistent with circular economy principles. The ECB's climate score methodology accounts for CO_2 emissions, carbon reduction targets, and environmental disclosures. The ECB intends to systematically reduce emissions from its institutional bond assets by reallocating purchases toward issuers with enhanced climate performance [69]. The Bank of England [70] integrates circular economy and resource efficiency into its climate monitoring frameworks and advances transparency under the Task Force on Climate-related Financial Disclosures. The People's Bank of China [71] operates an emissions reduction center, a green finance mechanism that provides concessional financing to commercial banking institutions for ecological credit.

4.12 GHG emissions from soybean production and clean energy alternatives

Figure 8 illustrates the relationship between soybean greenhouse gas (GHG) emissions and access to renewable energy from 2000 to 2022. During this period, soybean-related GHG emissions remained consistently low, below 1 kiloton (kt) per year, with only marginal variations. In contrast, access to renewable energy, comparably constrained in the early 2000s, began to increase gradually in 2009 and rose significantly after 2014, reaching 10 kt by 2022. Despite advances in energy infrastructure, there has been no substantial increase in soybean GHG emissions, indicating a decoupling effect between emission growth and energy development. This trend is consistent with the ecological modernization theory [47], which posits that technological advancements, such as renewable energy adoption, can enhance environmental performance without compromising production.

While soybean production does not constitute a primary driver of increased renewable energy availability, it may have contributed indirectly to emissions stabilization. Energy-efficient processing and storage methods utilizing renewable energy, such as solar dryers and biogas, can reduce the soybean value chain's

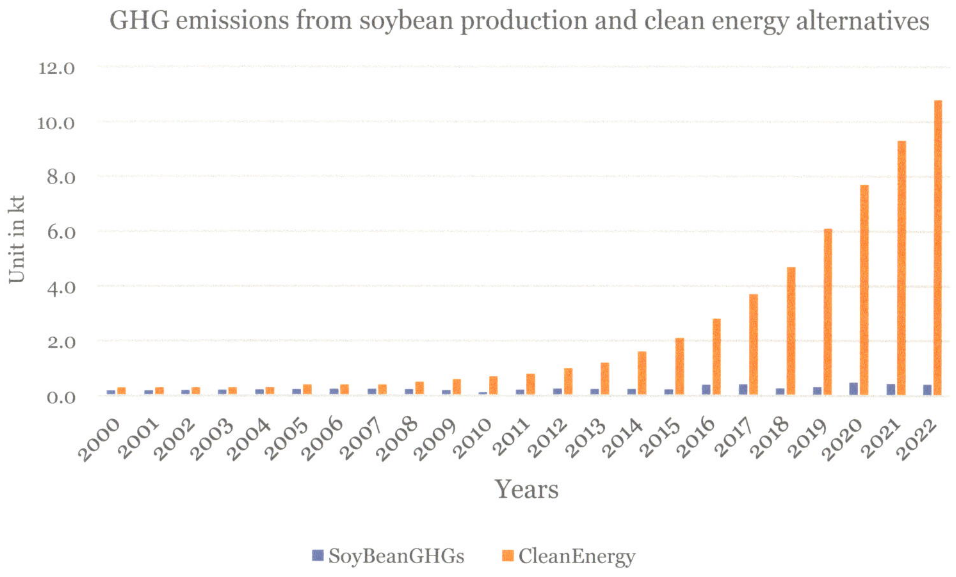

Figure 8.
GHG emissions from soybean production and clean energy alternatives. Source: authors' computation using data from [10, 11].

dependence on fossil fuels [72]. This is consistent with the sustainable development theory, which underscores the importance of clean technology in mitigating the environmental impacts of economic activity. The consistently low GHG emissions from soybean farming can be attributed to its relatively lower energy intensity compared with other crops [48]. Furthermore, soybeans contribute to nitrogen fixation in soils, thereby reducing the need for synthetic nitrogen fertilizers, a major source of GHG emissions in alternative crops. This finding illustrates a fundamental CSA principle [73]: sustained productivity, enhanced resilience through reduced emission intensity, and greater mitigation through the adoption of clean energy technologies. The soybean industry, while not a significant contributor to GHG emissions, derives advantages from the broader renewable energy transition, which contributes to a sustainable low-carbon profile.

4.13 Transportation and soybean trade (export and import)

Transport services (INFRST), used as a proxy for infrastructure, encompass the movement of soybeans across regions, together with ancillary services that facilitate and support logistics and trade efficiency. Transport is measured as the proportion of services, whereas SBEXP represents the soybean export quantity index, and SBIMP denotes the soybean import quantity index.

Figure 9 illustrates the relationship between transportation infrastructure and the soybean system. Transport infrastructure investment has remained moderate and relatively stable over the years, with modest increases in 2005–2007 and after 2015. This indicates that transportation infrastructure was unable to significantly improve despite variations in the soybean trade. A notable feature is the substantial increase in soybean exports in 2017, exceeding 1,600 tonnes. This is likely attributable to a temporary trade liberalization or policy shift, possibly

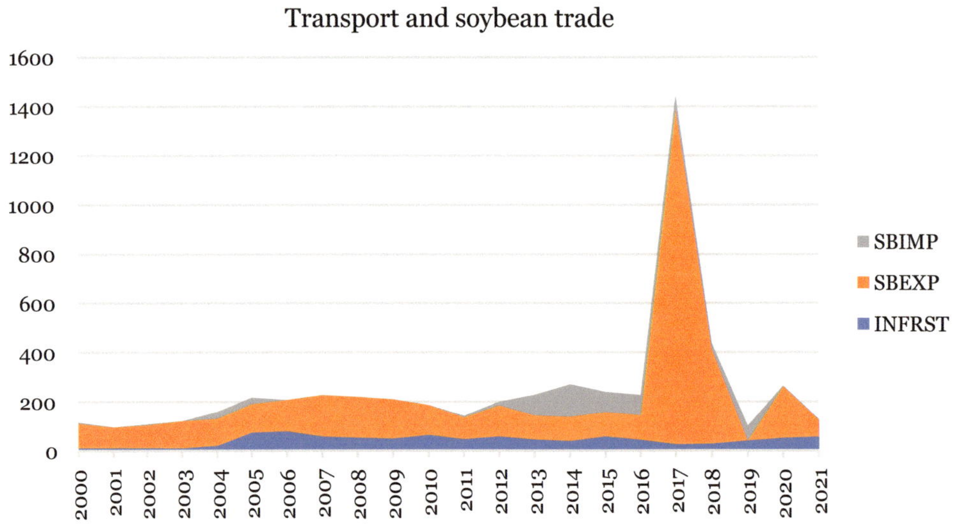

Figure 9.
Transport and soybean trade system. Source: Authors' computation using data from [10, 11].

driven by increased foreign demand or favorable trade agreements. Imports began to increase gradually between 2010 and 2015 and peaked just before 2017. A sharp decline occurred in 2018, possibly due to higher domestic production or export restrictions. According to [74, 75], export increases in SSA countries, including Nigeria, are often associated with global soybean demand, particularly from China and India, as well as short-term trade liberalization or currency devaluation, which enhances competitiveness. Nigerian oilseed exports were partly driven by temporary fluctuations in demand and exchange rate effects [75]. According to the World Bank [76], Nigeria's transportation system constitutes one of the weakest components in the agricultural value chain, with inadequate rural road connectivity raising farm-gate costs by 25–50% on average, thereby reducing soybean export competitiveness. This partly explains why exports did not continue beyond 2017.

Low yields (arising from low-quality seed varieties, suboptimal agronomic practices, and climatic stress) and inadequate irrigation systems continue to undermine Nigeria's goal of soybean self-reliance, resulting in substantial reliance on imports. For instance, in 2016, Nigeria imported approximately 31% of its total soybean demand, highlighting the gap between domestic production and consumption [77]. The soybean processing industry requires the strengthening of domestic demand, thereby reducing reliance on costly imports [78]. Achieving this requires institutional partnerships among governments, the private sector, and development banks to invest in soybean crushing infrastructure and bio-refineries, as well as the reallocation of underutilized land from traditional agriculture to alternative cash crops [79]. Since soybean crushing produces only 18% oil and 78% meal, oil-only processing is economically unviable in the absence of a local market for soybean cake. Due to limited crushing facilities, Ethiopia exported 53.94 thousand MT of soybeans and imported 4.52 thousand MT of soy oil in 2017 [80]. The 2017 increase may also be attributed to global price fluctuations or domestic responses to food security issues. UNCTAD [78] documented price increases in global soy markets, which often led to trade surges in low-income countries. Moreover, the stagnant trend in INFRST (transport) suggests that without additional investment in transport, storage, and rural roads, soybean export growth may be unsustainable. This aligns with [81], which emphasizes that global market accessibility for developing-country exports is constrained not only by trade barriers but also by domestic structural constraints such as insufficient storage, transportation, and logistics infrastructure. In commodity sectors such as soybeans, these barriers limit both export volumes and quality standards.

4.14 Irrigation practices in relation to soybean production

Figure 10 illustrates the relationship between irrigation (ldirrig), an indicator of agronomic integration, and soybean output. Soybean production generally increased, with minor fluctuations. Production rose steadily from 2000 to 2007, peaking in 2007, before declining sharply between 2008 and 2010. From 2011 to 2014, a rising trend began, leading to a major increase between 2015 and 2020, reaching its peak in 2020, followed by a slight decline in 2021–2022, although maintaining levels above prior years. Land under irrigation was comparatively lower in scale and largely stable over time, with only marginal increases, indicating slow expansion of irrigation infrastructure. Despite continuous or slightly increasing irrigation, substantial productivity gains (such as those recorded between 2015 and 2020) suggest that other factors, including government support, mechanization, improved seed varieties, or climatic conditions, may have played a more significant role. This aligns with FAOSTAT [82], which notes that insufficient irrigation coverage constrains stable soybean yields, particularly in Nigeria's rain-fed zones. The limited investment in irrigation does not correspond with growing demand or production objectives. This is consistent with UNCTAD's [81] findings that

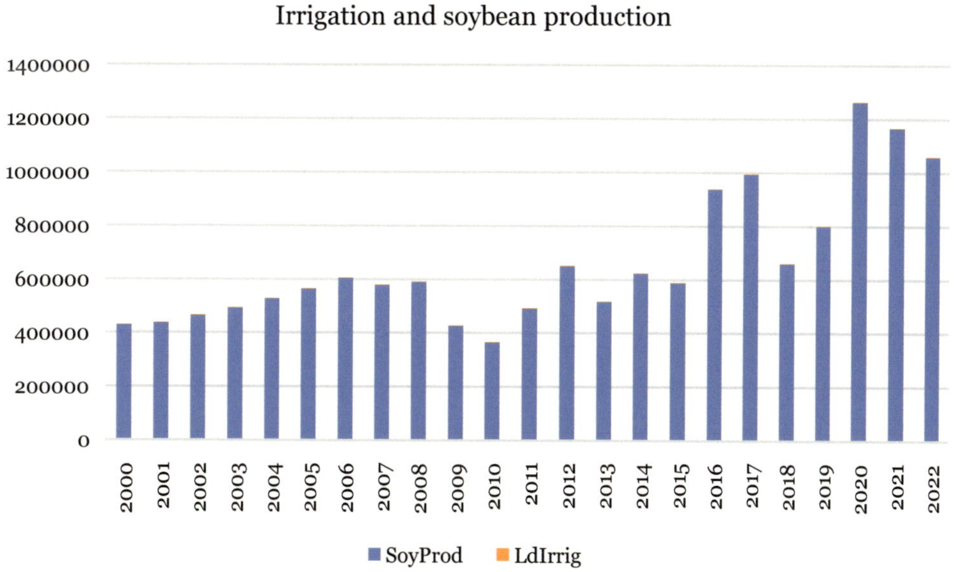

Figure 10.
Irrigation and soybean production. Source: authors' computation using data from [10, 11].

agricultural productivity improvements are unsustainable without complementary infrastructure such as irrigation and storage. The declines in 2009–2010 and again in 2021–2022 may be attributed to severe weather events or economic recessions (such as monetary crises and COVID-19), indicating vulnerability due to limited resilience within the production system.

According to FAO [83], irrigation accounts for only 1% of Nigeria's cultivated land, severely constraining year-round agricultural productivity, particularly for crops such as soybean that benefit from off-season planting. To enhance the climate resilience of soybean production systems, agronomic risk mitigation strategies involving irrigation infrastructure, agricultural insurance, and the adoption of more efficient inputs (such as drought-resistant cultivars and fertilizers) must be integrated, particularly in areas vulnerable to unpredictable rainfall and resource limitations [21, 22, 84]. If production increased without significant irrigation expansion, it suggests that yield per hectare improved, possibly due to CSA practices, improved seed varieties, soil management, optimized input use, or the cultivation of additional land, as highlighted by FAO [26, 75]. Consequently, irrigation development has lagged behind soybean production growth, thereby rendering the sector vulnerable to climatic variability. Expanding irrigation, alongside storage and transport infrastructure, is essential to sustaining and stabilizing gains in soybean productivity.

5. Conclusion and policy recommendations

Climate variability and change pose a significant risk to Nigerian soybean production, with variations in rainfall patterns, temperature extremes, and the increased frequency of extreme weather events adversely affecting yield stability and farmer incomes, as demonstrated through ARDL, correlation, and Granger causality tests. The trend analysis reveals that soybean output and CO_2 emissions exhibit synchronized movements, indicating that higher production activities contribute directly to increased CO_2 emissions. The ARDL results further show that CO_2 levels exert a modest but positive and marginally significant short-run effect on soybean output. Granger causality tests revealed that soybean production influences harvested area, temperature, and domestic credit. Soybeans, a crucial commodity in Nigeria's agricultural sector and an important protein source, are particularly vulnerable to climatic stresses, with direct implications for food security, nutrition, and economic sustainability. Ultimately, developing a resilient soybean sector through effective adaptation measures not only safeguards Nigeria's food system against climate-related risks but also advances broader development objectives, including sustainability, rural development, and national food self-sufficiency. A coordinated, multistakeholder effort is therefore essential to transform structural vulnerabilities into adaptive capacity and ensure the long-term viability of soybean production in a changing climate. The research recommends that:

i. Strengthen the adoption of CSA practices for soybean farming, including conservation agriculture, crop rotation, soil moisture management, and agroforestry systems, to improve resilience to climatic variability. Additionally, increase investment in research and development (R&D) to produce soybean varieties that are drought-resistant and heat-resistant. Enhance the capacity of agricultural organizations and seed distribution networks to ensure that farmers have access to, and effectively utilize, improved seed varieties.

ii. Improve local weather forecasting and climate information systems to provide farmers with timely and reliable data that supports informed decision-making regarding planting and harvesting schedules under variable climatic conditions. Develop affordable, decentralized irrigation systems, and promote rainwater harvesting as strategies to reduce reliance on variable rainfall. Implement measures to enhance the accessibility of such technologies, particularly for smallholder farmers.

iii. Expand access to climate risk insurance, microcredit, and subsidies to safeguard farmers against crop losses and encourage the adoption of resilient agricultural practices. Enhanced insurance coverage could contribute to income stabilization, stimulate agricultural production investment, and strengthen resilience to climate-related risks. Furthermore, promote the adoption of land conservation practices, reforestation, and soil fertility enhancement to sustain agricultural ecosystems and mitigate environmental degradation associated with climatic variability.

iv. Increased soybean production has contributed to improved food supply and economic gains; however, it is associated with significant environmental consequences, particularly in CO_2 emissions. Sustainable farming practices (e.g., precision agriculture, conservation tillage) require strong policy support, the adoption of climate-smart technologies to reduce emissions intensity, and monitoring and reporting systems to track and mitigate agricultural emissions. Furthermore, strengthening national and regional policies that integrate climate adaptation into agricultural planning, alongside fostering public–private partnerships to facilitate innovation, mobilize funding, and enhance infrastructure development, is essential for sustainable soybean production.

v. Establish climate-smart financing mechanisms and green finance instruments to support the scaling up of clean energy adoption. Introduce subsidies, tax incentives, and targeted grants for renewable energy technologies tailored to farmer organizations and agro-processors, particularly in rural areas.

vi. To promote the integration of clean energy in soybean production systems, the government should expand subsidies and tax incentives for solar-powered agricultural equipment, establish and strengthen solar-powered agro-processing cooperatives in major soybean-producing areas, and enhance partnerships with clean energy service providers to design accessible and inclusive financing schemes for farmers. Incorporating clean energy measures into Nigeria's National Agricultural Technology and Innovation Plan (NATIP), alongside optimizing irrigation, improving postharvest storage, and strengthening climate resilience, could be critical to reducing import dependence while sustaining export performance.

Acknowledgments

We acknowledge the Food and Agriculture Organisation (FAO) and the World Bank through the World Development Indicators (WDI) and Climate Knowledge Portal for making the data available for use.

Conflict of Interest

The authors declare no conflict of interest.

Funding

The author received no funds for this research.

Ethical consideration

The research relies on publicly accessible secondary datasets, which help to avoid any ethical concerns. The researchers make sure to properly credit all data sources to maintain transparency and uphold academic integrity.

Author details

Adeyinka Richard Aroyehun[1*], Adanna Henri-Ukoha[1] and Elizabeth Ekaette Oladejo[2]

1 Agricultural Economics and Agribusiness Management Department, University of Port Harcourt, Choba, Port Harcourt, Rivers State, Nigeria

2 Fortbend Bend Independent School District, Sugar Land, Texas, USA

*Address all correspondence to: aroyehun_adeyinka@uniport.edu.ng

References

[1] UNICEF-WHO-The World Bank Group. Joint child malnutrition estimates- 2017 edition [Internet]. 2017. Available from: https://data.unicef.org/resources/joint-childmalnutrition-estimates-2017-edition/ [Accessd: 2025-February-12]

[2] Food and Agriculture Organization (FAO). Smallholders and family farmers [Internet]. 2012. Available from: https://www.fao.org/fileadmin/templates/nr/sustainability_pathways/docs/Factsheet_SMALLHOLDERS.pdf [Accessed: 2025-March-02]

[3] Foyer CH, Siddique KHM, Tai APK, Anders S, Fodor N, Wong F, Lam H. Modelling predicts that soybean is poised to dominate crop production across Africa. Plant, Cell & Environment. 2019;**42**(1):373–385. DOI: 10.1111/pce.13466

[4] Anderson W, Johansen C, Siddique KHM. Addressing the yield gap in rainfed crops: A review. Agronomy for Sustainable Development. 2016;**36**:18. DOI: 10.1007/s13593-015-0341-y

[5] Mauser W, Klepper G, Zabel F, Delzeit R, Hank T, Putzenlechner B, Calzadilla A. Global biomass production potentials exceed expected future demand without the need for cropland expansion. Nature Communications. 2015;**6**:8946. DOI: 10.1038/ncomms9946

[6] Idumah FO, Mangodo C, Ighodaro UB, Owombo PT. Climate change and food production in Nigeria: Implications for food security in Nigeria. Journal of Agricultural Science. 2016;**8**(2):73–83. DOI: 10.5539/jas.v8n2p74

[7] Khojely DM, Ibrahim SE, Sapey E, Han T. History, current status, and prospects of soybean production and research in sub-Saharan Africa. The Crop Journal. 2018;**6**(3):226–235. DOI: 10.1016/j.cj.2018.03.006

[8] International Institute of Tropical Agriculture (IITA). Transforming African Agriculture through Research: IITA 2016 Annual Report. Ibadan: IITA; 2017. Available from: https://www.iita.org/wp-content/uploads/2017/08/2016-annual-reportLOWRES.pdf

[9] Thierfelder C, Matemba-Mutasa R, Rusinamhodzi L. Yield response of maize (Zea mays L.) to conservation agriculture cropping system in Southern Africa. Soil and Tillage Research. 2015;**146**:228–242. DOI: 10.1016/j.still.2014.10.002

[10] Food and Agriculture Organization (FAO). FAOSTAT Statistical Database [Internet]. Rome, Italy. 2025. Available from: https://www.fao.org/faostat [Accessed: 2025-January-13]

[11] World Bank. World Development Indicators [Internet]. Washington, DC: The World Bank. 2025. Available from: https://databank.worldbank.org/source/world-development-indicators [Accessed: 2025-January-15]

[12] World Bank Group. Climate Change Knowledge Portal (CCKP) [Internet]. Washington, DC: The World Bank. 2025. Available from: https://climateknowledgeportal.worldbank.org [Accessed: 2025-January-15]

[13] Vanlauwe B, Bationo A, Chianu J, Giller KE, Merckx R, Mokwunye U, et al. Integrated soil fertility management: Operational definition and consequences for implementation and dissemination. Outlook on

Agriculture. 2010;**39**(1):17–24. DOI: 10.5367/000000010791169998

[14] Chikowo R, Mapfumo P, Nyamugafata P, Giller KE. Maize productivity and nutrient and water use efficiencies across soil fertility domains on smallholder farms in Zimbabwe. Field Crops Research. 2014;**164**:136–147. DOI: 10.1016/j.fcr.2014.05.015

[15] Kassie M, Jaleta M, Shiferaw B, Mmbando F, Mekuria M. Adoption of interrelated sustainable agricultural practices in smallholder systems: Evidence from rural Tanzania. Technological Forecasting and Social Change. 2015;**80**(3):525–540. DOI: 10.1016/j.techfore.2012.08.007

[16] Bacci M, Ousman Baoua Y, Tarchiani V. Agrometeorological forecast for smallholder farmers: A powerful tool for weather-informed crops management in the Sahel. Sustainability. 2020;**12**(8):3246. DOI: 10.3390/su12083246

[17] Ajeigbe HA, Abdoulaye T, Chikoye D Improved crop–livestock system for enhanced food security and income generation in West Africa. 2010:60. Ibadan: IITA. Available from: https://hdl.handle.net/10568/90302 [Accessed: 2025-January-12]

[18] Abate T, van Huis A, Ampofo JKO. Pest management strategies in traditional agriculture: An African perspective. Annual Review of Entomology. 2014;**45**(1):631–659. DOI: 10.1146/annurev.ento.45.1.631

[19] Greatrex H, Hansen JW, Garvin S, Diro R, Blakeley S, Le Guen M, et al. Scaling up index insurance for smallholder farmers: Recent evidence and insights. CCAFS Working Paper. 2015. Available from: https://hdl.handle.net/10568/53101 [Accessed: 2025-March-07]

[20] Deressa TT, Hassan RM, Ringler C, Alemu T, Yesuf M. Determinants of farmers' choice of adaptation methods to climate change in the Nile Basin of Ethiopia. Global Environmental Change. 2009;**19**(2):248–255. DOI: 10.1016/j.gloenvcha.2009.01.002

[21] Hazell P. Potential role for insurance in managing catastrophic risk in developing countries. IFPRI. 2011. Available from: https://farm-d.org/wp-content/uploads/2024/03/Insurance-in-Managing-Catastrphic-Risks-in-DCs.pdf [Accessed: 2025-February-20]

[22] Nigeria Incentive Based Risk Sharing System for Agricultural Lending (NIRSAL). Newsletter: CBN's LDR policy a major boost for agriculture finance. Abuja: NIRSAL. 2020. Available from: https://nirsal.com/wp-content/uploads/2020/03/NIRSAL-Newsletter-_-Q1-2020.pdf [Accessed: 2025-January-25]

[23] World Bank. Transforming agribusiness for inclusive recovery, job creation, and poverty reduction (Report No. 35589). Washington, DC: World Bank; 2021. Available from: https://openknowledge.worldbank.org/handle/10986/35589 [Accessed: 2025-January-12]

[24] Karlan D, Savonitto B, Thuysbaert B, Udry C. Impact of savings groups on the lives of the poor. Proceedings of the National Academy of Sciences of the United States of America. 2014;**111**(12):4147–4151. DOI: 10.1073/pnas.1308044111

[25] GSMA. Digital Agriculture Maps 2020: State of the Sector in Low and Middle-Income Countries. London:

GSMA; 2020. Available from: https://www.gsma.com/r/wp-content/uploads/2020/10/GSMA-Agritech-Digital-Agriculture-Maps-2020-1.pdf

[26] Food and Agriculture Organization (FAO). Climate-smart agriculture sourcebook. Rome: FAO; 2013. Available from: https://www.fao.org/docrep/018/i3325e/i3325e.pdf [Accessed: 2025-January-12]

[27] Rockström J, Karlberg L, Wani SP, Barron J, Hatibu N, Oweis T, et al. Managing water in rainfed agriculture: The need for a paradigm shift. Agricultural Water Management. 2010;**97**(4):543–550. DOI: 10.1016/j.agwat.2009.09.009

[28] World Bank. Unlocking the potential of agriculture for inclusive growth and job creation in Nigeria. Washington, DC: World Bank; 2021. Available from: https://documents.worldbank.org [Accessed: 2025-January-15]

[29] Kaminski J, Christiaensen L. Post-harvest loss in sub-Saharan Africa—what do farmers say? Global Food Security. 2014;**3**(3–4):149–158. DOI: 10.1016/j.gfs.2014.10.002

[30] Federal Ministry of Agriculture and Rural Development. National agricultural technology and innovation policy (NATIP) 2022–2027. 2022. Available from: https://faolex.fao.org/docs/pdf/nig214137.pdf [Accessed: 2025-February-20]

[31] Sultan B, Gaetani M. Agriculture in West Africa in the twenty-first century: Climate change and impact scenarios, and potential for adaptation. Frontiers in Plant Science. 2016;**7**:1262. DOI: 10.3389/fpls.2016.01262

[32] International Renewable Energy Agency (IRENA) and Food and Agriculture Organization of the United Nations (FAO). Renewable energy for agri-food systems – Towards the Sustainable Development Goals and the Paris agreement. Abu Dhabi and Rome; 2021. DOI: 10.4060/cb7433en

[33] United States Agency for International Development Nigeria (USAID/Nigeria). Country Development Cooperation Strategy 2015-2020. 2015. Available from: http://gcan.ifpri.info/files/2017/03/Nigeria_CDCS_2015-2020.pdf [Accessed: 2025-January-12]

[34] International Renewable Energy Agency (IRENA). Renewable Energy Benefits: Decentralised Solutions in the Agri-food Chain. Abu Dhabi: IRENA; 2016. Available from: https://www.irena.org/publications/2016/Sep/Renewable-Energy-Benefits-Decentralised-solutions-in-agri-food-chain

[35] Sustainable Energy for All (SEforALL). Nigeria's integrated energy planning for universal access. SEforALL. 2021. Available from: https://nigeria-iep.sdg7energyplanning.org/ [Accessed: 2025-February-23]

[36] Federal Ministry of Environment. Nigeria's updated Nationally Determined Contributions (NDCs). Abuja: Government of Nigeria. 2021. Available from: https://www.environewsnigeria.com/nigeria-submits-final-report-of-updated-ndcs-to-unfccc/ [Accessed: 2025-April-10]

[37] Global Donor Platform for Rural Development. Financing food systems transformation and rural revitalization: Opportunities and challenges. 2024 Annual General Assembly Background

Paper. 2024. Available from: https://www.donorplatform.org/wp-content/uploads/2025/01/GDPRD-Draft-Background-Paper-for-AGA2024_2024NOV19.pdf [Accessed: 2025-February-20]

[38] Intergovernmental Panel on Climate Change IPCC. Climate change 2021: The physical science basis. Contribution of Working Group I to the Sixth Assessment Report of the IPCC [Internet]. Cambridge University Press. 2021. Available from: https://www.ipcc.ch/report/sixth-assessment-report-working-group-i/ [Accessed: 2025-April-10]

[39] Ayanlade A, Radeny M, Morton JF. Comparing smallholder farmers' perception of climate change with meteorological data: A case study from southwestern Nigeria. Weather and Climate Extremes. 2017;**15**:24–33. DOI: 10.1016/j.wace.2016.12.001

[40] Lobell DB, Gourdji SM. The influence of climate change on global crop productivity. Plant Physiology. 2012;**160**(4):1686–1697

[41] Tendall DM, Joerin J, Kopainsky B, Edwards P, Shreck A, Le QB, Kruetli P, Grant M, Six J. Food system resilience: Defining the concept. Global Food Security. 2015;**6**:17–23. DOI: 10.1016/j.gfs.2015.08.001

[42] Ingram J. A food systems approach to researching food security and its interactions with global environmental change. Food Security. 2011;**3**(4):417–431

[43] Coulibaly A, Motelica-Heino M, Hien E. Determinants of agroecological practices adoption in the Sudano-Sahelian Zone. Journal of Environmental Protection. 2019;**10**:900–918. DOI: 10.4236/jep.2019.107053

[44] Pretty J. Agricultural sustainability: Concepts, principles and evidence. Philosophical Transactions of the Royal Society B: Biological Sciences. 2008;**363**(1491):447–465

[45] Reinesch A, Fausak L, Joseph A, Lavkulich L. Water, energy, and nutrient losses from food wastage of selected crops in three agro-climatic zones in British Columbia, Canada. Agricultural Sciences. 2022;**13**:947–972. DOI: 10.4236/as2022.138059

[46] Cai J, Ung L, Setboonsarng S, Leung P Rice contract farming in Cambodia: Empowering farmers to move beyond the contract toward Independence. ADBI Working Paper. [Internet]. 2008. Available from: https://www.adb.org/sites/default/files/publication/156748/adbi-dp109.pdf [Accessed: 2025-April-10]

[47] Mol APJ, Sonnenfeld DA. Ecological modernisation around the world: Perspectives and critical debates. Routledge. Economic Geography. 2000;**77**(4). DOI: 10.2307/3594110

[48] Hill J, Nelson E, Tilman D, Polasky S, Tiffany D. Environmental, economic, and energetic costs and benefits of biodiesel and ethanol biofuels. Proceedings of the National Academy of Sciences. 2006;**103**(30):11206–11210. DOI: 10.1073/pnas.0604600103

[49] Pesaran MH, Shin Y, Smith RJ. Bounds testing approaches to the analysis of level relationships. Journal of Applied Econometrics. 2001;**16**:289–326. DOI: 10.1002/jae.616

[50] Ntiamoah EB, Li D, Appiah-Otoo I, Twumasi MA, Yeboah EN. Towards a sustainable food production: Modelling the impacts of climate change on maize and soybean production in Ghana. Environmental Science and Pollution Research. 2022;**29**:72777–72796. DOI: 10.1007/s11356-022-20962-z

[51] Phillips PCB, Perron P. Testing for a unit root in time series regression. Biometrika. 1988;**75**:335–346. DOI: 10.1093/biomet/75.2.335 [Accessed: 2025-March-20]

[52] Dickey DA, Fuller WA. Distribution of the estimators for autoregressive time series with a unit root. Journal of the American Statistical Association. 1979;**74**:427–431. DOI: 10.1080/01621459.1979.10482531

[53] Kucharik CJ, Serbin SP. Impacts of recent climate change on Wisconsin corn and soybean yield trends. Environmental Research Letters. 2008;3. DOI: 10.1088/1748-9326/3/3/034003

[54] Jiang Q, Qi Z, Xue L, Bukovsky M, Madramootoo CA, Smith W. Assessing climate change impacts on greenhouse gas emissions, N losses in drainage, and crop production in a subsurface drained field. Science of the Total Environment. 2020;**705**:135969. DOI: 10.1016/j.scitotenv.2019.135969

[55] Rehman A, Ma H, Ozturk I. Decoupling the climatic and carbon dioxide emission influence to maize crop production in Pakistan. Air Quality, Atmosphere & Health. 2020;**13**:695–707. DOI: 10.1007/s11869-020-00825-7

[56] Rehman A, Ma H, Ahmad M, Ozturk I, Chishti MZ. How do climatic change, cereal crops and livestock production interact with carbon emissions? Updated evidence from China. Environmental Science and Pollution Research. 2021;**28**(24):30702–30713. DOI: 10.1007/s11356-021-12948-0

[57] Warsame AA, Sheik-Ali IA, Ali AO, Sarkodie SA. Climate change and crop production nexus in Somalia: An empirical evidence from ARDL technique. Environmental Science and Pollution Research. 2021;**28**:19838–19850. DOI: 10.1007/s11356-020-11739-3

[58] Zhang W, Qian C, Carlson KM, Ge X, Wang X, Chen X. Increasing farm size to improve energy use efficiency and sustainability in maize production. Food and Energy Security. 2021;**10**. DOI: 10.1002/fes3.271

[59] Feng Y, Zhang Y, Li S, Wang C, Yin X, Chu Q, Chen F. Sustainable options for reducing carbon inputs and improving the eco-efficiency of smallholder wheat-maize cropping systems in the Huanghuaihai Farming Region of China. Journal of Cleaner Production. 2020; **244**(26):118887. DOI: 10.1016/j.jclepro.2019.118887

[60] Chen R, Zhang R, Han H, Jiang Z. Is farmers' agricultural production a carbon sink or source? – Variable system boundary and household survey data. Journal of Cleaner Production. 2020;**266**. DOI: 10.1016/j.jclepro.2020.122108

[61] Liu D, Zhang W, Wang X, Guo Y, Chen X. Greenhouse gas emissions and mitigation potential of hybrid maize seed production in northwestern China. Environmental Science and Pollution Research. 2022;**29**:17787–17798. DOI: 10.1007/s11356-021-16990-w

[62] Arrieta EM, Cuchietti A, Cabrol D, Gonzalez AD. Greenhouse gas emissions

and energy efficiencies for soybeans and maize cultivated in different agronomic zones: A case study of Argentina. Science of the Total Environment. 2018;**625**:199–208. DOI: 10.1016/j.scitotenv.2017.12.286

[63] Kukal MS, Irmak S. Climate-driven crop yield and yield variability and climate change impacts on the U.S. Great Plains agricultural production. Scientific Reports. 2018;8. DOI: 10.1038/s41598-018-21848-2

[64] Okpukpara BC, Adebayo O, Ukwuaba IC. Smallholder farmers' access to agricultural insurance schemes: An analysis of the inhibitors in Kogi State, Nigeria. Turkish Journal of Agriculture - Food Science and Technology. 2021;**9**(12):2159–2165. DOI: 10.24925/turjaf.v9i12.2159-2165.4381

[65] Aroyehun AR. Determinants of poultry egg farmers' participation in livestock insurance in Rivers State, Nigeria. Black Sea Journal of Agriculture. 2023;**6**(6):643–648. DOI: 10.47115/bsagriculture.1323005

[66] Henri-Ukoha A, Ndubueze-Ogaraku ME, Obi SE, Aroyehun AR. Willingness to pay for flood insurance by cassava-based farmers in Ikwerre Local Government Area, Rivers State, Nigeria. In: Onyeneke RU, Emenekwe CC, Nwajiuba CU, editors. Energy Transition, Climate Action and Sustainable Agriculture. Cham: Springer; 2025. p. 513–526. DOI: 10.1007/978-3-031-83165-2_25

[67] National Agricultural Insurance Commission (NAIC). Stronger and more resilient. NAIC 2020 Annual report. National Agricultural Insurance Commission, Nigeria. 2021. Available from: https://content.naic.org/sites/default/files/annual_report_2020.pdf [Accessed: 2025-March-15]

[68] Ifeanyi-Obi CC, Etuk UR, Jike-Wai O. Climate change, effects and adaptation strategies; Implications for agricultural extension system in Nigeria. Greener Journal of Agricultural Sciences. 2017;**3**(6):503–507

[69] European Central Bank. Guide on climate-related and environmental risks: Supervisory expectations relating to risk management and disclosure [Internet]. Frankfurt am Main: European Central Bank; 2020 Nov Available from: https://www.bankingsupervision.europa.eu/press/pr/date/2020/html/ssm.pr201127~5642b6e68d.en.html [Accessed: 2025-July-22]

[70] Bank of England. Climate Biennial Exploratory Scenario (CBES) [Internet]. London: Bank of England; 2022 May 24 Available from: https://www.bankofengland.co.uk/stress-testing/2022/results-of-the-2021-climate-biennial-exploratory-scenario [Accessed: 2025-July-22]

[71] Economic Insights Global Economic and Markets Research. Green central bank policy: People's Bank of China (PBoC). Sustainable Economics. 2021. Available from: https://www.commbankresearch.com.au/sfc/servlet.shepherd/version/renditionDownload?versionId=0684y000009OJi8AAG&rendition=ORIGINAL_PDF [Accessed: 2025-February-20]

[72] FAO. Climate-smart Agriculture and the Sustainable Development Goals: Mapping Interlinkages, Synergies and Trade-offs and Guidelines for Integrated Implementation. Rome; 2019 [Accessed: 2025-February-12]

[73] Lipper L, Thornton P, Campbell BM, Baedeker T, Braimoh A, Bwalya T, Torquebiau EF. Climate-smart agriculture for food security. Nature Climate Change. 2014;**4**: 1068–1072. DOI: 10.1038/nclimate2437

[74] Organization for Economic Co-operation and Development, Food and Agriculture Organization of the United Nations. OECD-FAO Agricultural Outlook 2019–2028: Africa Chapter. Paris/Rome: OECD & FAO; 2019. DOI: 10.1787/agr_outlook-2019-en

[75] Food and Agriculture Organization. Irrigation in Africa in Figures – AQUASTAT Survey 2019. FAO Water Reports No. 42. Rome: FAO; 2020. Available from: https://www.fao.org/3/a0232e/A0232E.pdf [Accessed: 2025-April-15]

[76] World Bank Group. Enabling the Business of Agriculture 2017: Nigeria country profile. Washington, DC: World Bank. 2017. Available from: https://hdl.handle.net/10986/25966 [Accessed: 2025-January-23]

[77] FAOSTAT. Database: Nigeria crop production and trade statistics. Rome: FAO; 2021 July. Available from: https://www.fao.org/faostat/ [Accessed: 2025-January-19]

[78] Sedibe MM, Mofokeng AM, Masvodza DR. Soybean production, constraints, and future prospects in poorer countries: A review. In: Hasanuzzaman M, editor. Production and Utilization of Legumes – Progress and Prospects. London: IntechOpen; 2023. DOI: 10.5772/interchpoen.109516 [Accessed: 2025-April-26]

[79] Debnath D, Babu S. Prospects for sustainable intensification of soybean production in Sub-Saharan Africa. African Journal of Agricultural and Resource Economics. 2020;**15** (4):365–371

[80] FAOSTAT. Databases: Soybean production and trade data. 2019. Available from: https://www.fao.org/faostat/en/#data [Accessed: 2025-July]

[81] UNCTAD. Commodities and Development Report: Market Trends. Geneva: United Nations Conference on Trade and Development; 2018. Available from: https://unctad.org/webflyer/commodities-and-development-report-2017

[82] FAOSTAT. Irrigation and crop production statistics. Rome: FAO; 2021. Available from: https://www.fao.org/faostat [Accessed: 2025-April-02]

[83] FAO. FAOSTAT country profile: Nigeria. 2020. Available from: https://www.fao.org/faostat/en/#country/159 [Accessed: 2025-April-2]

[84] Mghenyi E, Dankers C, Thurlow J, Anyiro C. Transforming Agribusiness in Nigeria for Inclusive Recovery, Jobs Creation, and Poverty Reduction: Policy Reforms and Investment Priorities. Washington, DC: International Development in Focus, World Bank; 2022. DOI: 10.1596/978-1-4648-1724-3

Section 5

Carbon Management and Planetary Systems

Chapter 6

Tele-illumination to Re-balance the Earth's Natural Carbon Cycle? A Review of the Large-scale Ocean Carbon Cycle

John Allen, Calum Fitzgerald and Lonnie Franks

Abstract

A new Nature Based Carbon Cycle Management Solution (NBCCMS) for capturing the entire man-made emission of carbon dioxide per year and locking it away in the deep ocean, called ECOPIA™ (Earth Climate Optimization Productivity Island Array), has been devised by a marine think tank, MyOcean Resources Ltd. Empowering the natural primary production capacity of the oceans solely through the provision of light at depth, tele-illumination, the proposed solution removes the excess atmospheric CO_2, de-acidifies the ocean's waters, creates new sustainable fisheries, and allows all world economies to continue to grow and prosper. The solution will affect the natural capture and storage of carbon, enabling the control and regulation of CO_2 levels in the atmosphere via natural mechanisms. It is a modular, scalable solution located in the oligotrophic sub-tropical gyres of the world's oceans. These otherwise minimally productive gyres are growing at a rate of 800,000 km^2 per year, at the cost of productive ocean areas. Simply providing light below the surface layers of these gyres, and nothing more, the solution provides no mechanism for preferential pressure on naturally determined biodiversity. Taking the, perhaps pessimistic, view that the world as a whole can only achieve static anthropogenic fossil fuels usage by the year 2030, then tele-illuminating <0.002% of the world's oceans will be required to return atmospheric CO_2 levels to 350 ppm. In this paper, we review the proposed solution within the framework of what is known about large-scale oceanic net primary production, more specifically export production.

Keywords: nature based solutions, climate change, ocean productivity, climate control, climostat, blue carbon, natural carbon sequestration

1. Introduction

We have become so accustomed to being instructed that there is no "silver bullet" to the anthropogenic climate crisis that most of us have begun to accept it as an irrefutable fact. However, there are no published papers demonstrating this, if indeed it is something that could be demonstrated. MyOcean Resources Ltd. believes this conviction results simply from a limit in our ability to comprehend the scale and capacity of our planet. For many, the ocean is still just the shallow coastal and shelf-edge waters lapping against the fringe of the Earth's continents and islands. Yet the global oceans cover more than 2/3 of our planet's surface, over 350,000,000 km^2, most of which is greater than 1 km deep, and much is greater than 4 km deep.

The characteristics and solar position of the Earth provide it with some important chemical cycles for life, not least among which is the water cycle, where evaporation at the Earth's surface is a source of water vapor for the atmosphere, and precipitation returns this back to the Earth's surface. A second critical cycle is the carbon cycle, where carbohydrate and hydrocarbon structures produce carbon dioxide (CO_2) through respiration and combustion just below or at the Earth's surface, which is released into the atmosphere and then taken up by biological primary production through photosynthesis and converted back into carbohydrates and hydrocarbons. We mention both of these natural cycles because it is highly likely that the latter would not exist without the former; indeed, they may both be co-dependent. However, it is the latter, or rather the anthropogenic exploitation of it, which is the subject of this review manuscript.

Considerably less than 0.5% of the Earth's combined carbon storage is naturally and actively cycling through respiration, combustion, and photosynthesis (**Figure 1**). There is a growing consensus that this carbon cycle has a natural balance in carbon mass of around 210 × 10^9 tonnes (billion tonnes, or Giga-tonnes [Gt], or peta-grammes [Pg]) annually (**Figure 1**). This cycle is relatively evenly divided between the oceans (~70% of the Earth's surface area) and the land, at ≥90–100 Gt each; although most estimates would put the land component up to 30% greater than the ocean [1, 2, 3] (to cite a few examples). One difficulty in estimating the size of this "natural" carbon cycle is that there is a very long history of anthropogenic deforestation, and fossil fuel burning may only have overtaken deforestation as an anthropogenic impact on this cycle in the latter half of the 20th century [2]. A second difficulty lies in the fact that nature has already begun to adjust for the anthropogenic imbalance; this will be discussed later. We note here that throughout this paper, we refer to carbon mass so that we have continuity regardless of whether we are referring to CO_2 or to more complex carbohydrate or hydrocarbon structures.

As a result of, and possibly partly a cause of, both a once exponentially growing global population [3] and an unprecedented rate of innovation, the burning of fossil fuels has been increasing at an exponentially growing rate over the last 150–250 years. The result is an imbalance favoring the combustion side of the carbon cycle (**Figure 2**). This anthropogenic imbalance in the Earth's carbon cycle, ~ 9 Gt, only represents an approximately 5% increase in the combustion side of the Earth's carbon cycle, but the Earth is having difficulty re-balancing the system. All the effects of this imbalance, on global temperature, on ocean acidification, on the cryosphere, and on ocean circulation, have been shown to have positive feedback consequences [4, 5]. The result of solely positive feedback is that these cumulative effects are getting more extreme with time and follow the exponential relationship in anthropogenic carbon emissions.

DOI: http://dx.doi.org/10.5772/intechopen.1014740

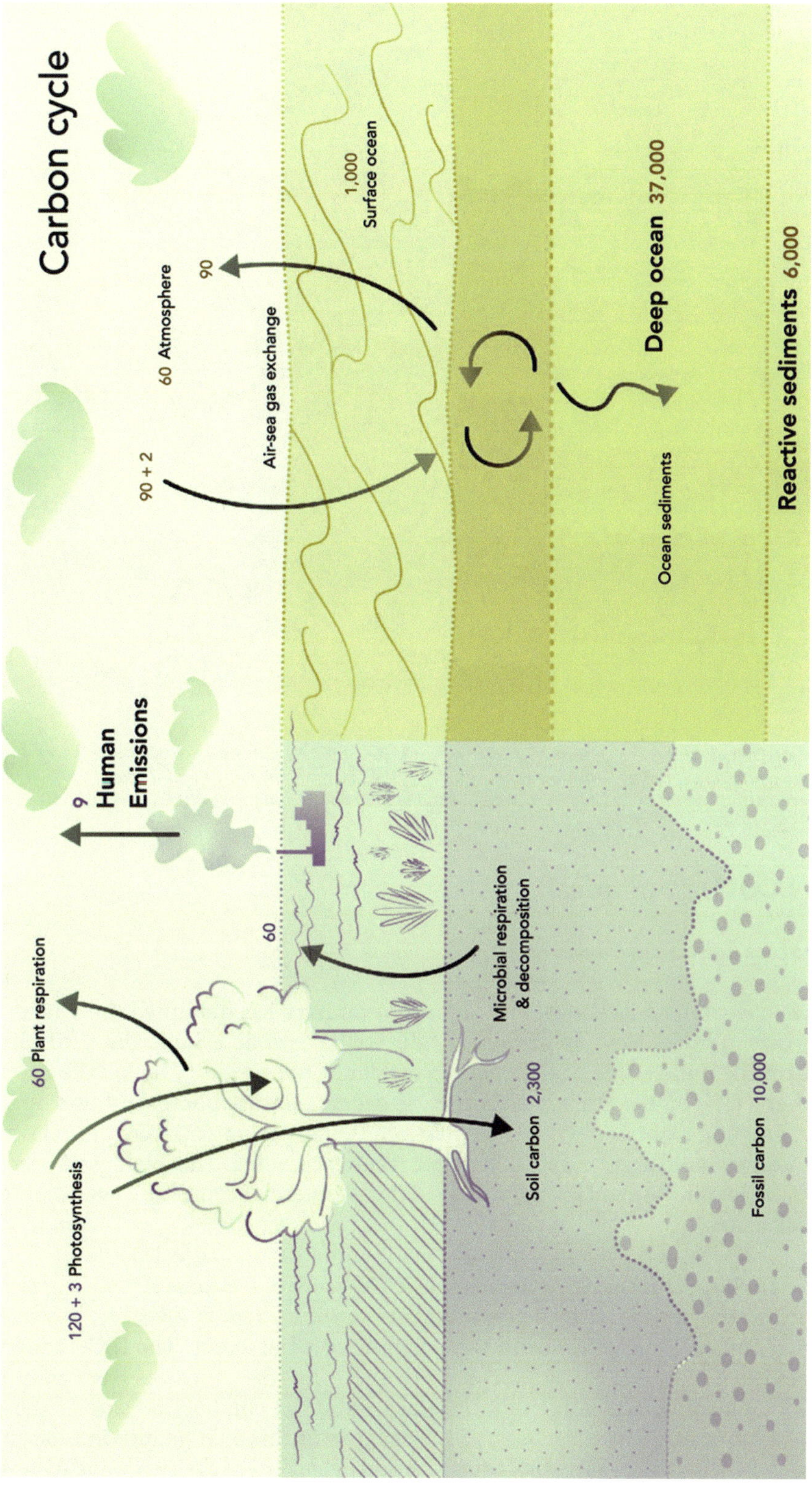

Figure 1.
Earth's natural carbon cycle, adapted from U.S. Department of Energy, Biological and Environmental Research Information System [1], (numbers in Gigatonnes of Carbon).

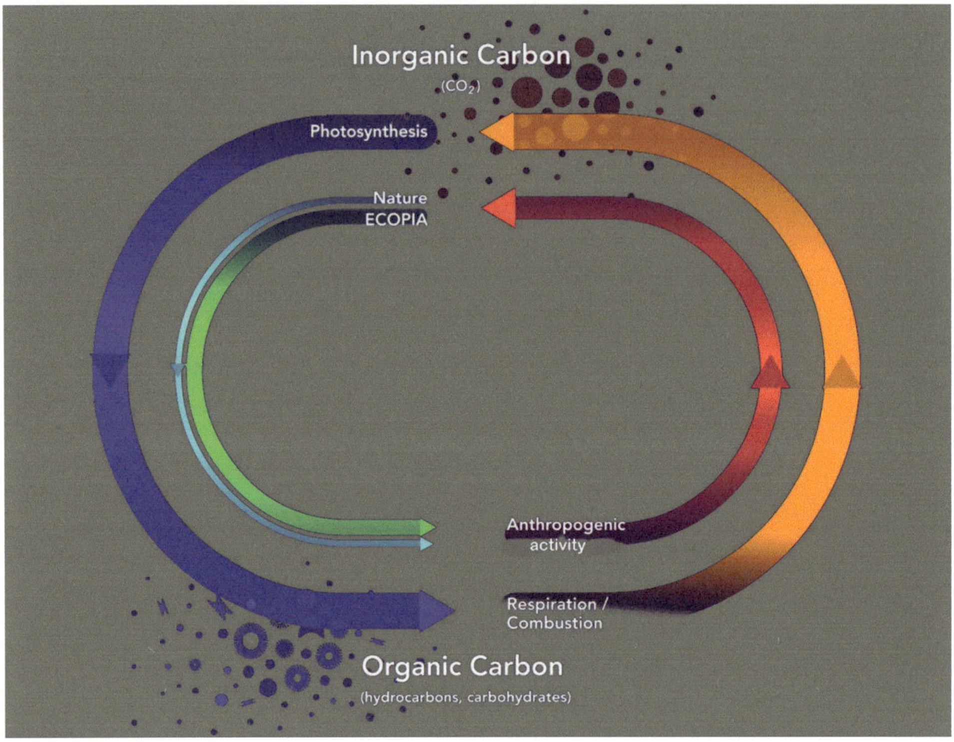

Figure 2.
Simplified cartoon of the Earth's carbon cycle: the background (~210 Gt) natural photosynthesis (blue) and respiration/combustion (orange), anthropogenic activity (~9 Gt in red), the natural adjustment (2–3 Gt) to anthropogenic rising atmospheric CO_2 (light blue), and the proposed re-balancing of ECOPIA™ (green).

2. The ECOPIA™ proposal

2.1 A nature-based approach

As we have described above, anthropogenic activity has disrupted the Earth's current carbon cycle. Now, in principle, "all" we have to do is build the natural booster for the uptake side of the cycle to rebalance the system. The uptake of atmospheric CO_2 on the Earth's carbon cycle occurs because photosynthesis by healthy, growing plant life exceeds respiration, thereby taking up CO_2 and releasing oxygen as they build their cellulose hydro-carbon structures. This biological process has been happening and adapting over geological timescales since the evolution of marine algae [6–8].

The Earth's carbon cycle has already made some natural adjustments in response to the currently rising atmospheric CO_2 concentrations. According to the latest IPCC report, the Earth's natural adjustment is as high as 5 Gt [9]; however, this figure is in the purest terms of the meaning of adjustment. The IPCC report expresses the magnitude to which the Earth's CO_2 uptake processes have adapted to the ~ 9 Gt loading that humankind has placed on the combustion side of the natural carbon cycle. Unfortunately, 50% or more of this 5 Gt adaptation has nothing to do with the natural biological carbon cycle. There are two ways in which atmospheric CO_2 can be taken up by the Earth. The first is known as the biological pump and is driven by photosynthesis; this is the process of synthesizing CO_2 molecules into carbohydrates and hydrocarbons and is the return side of the Earth's natural carbon cycle discussed above. The second is the solubility pump, which simply sets the background partitioning of CO_2. Increased atmospheric CO_2 concentration increases its solubility in rainwater, freshwater systems, and seawater, creating, in simple terms, increased carbonic acid in these aqueous reservoirs and hence a lower pH, i.e., acidified oceans and rainfall [10]. This is not a good thing and feeds back positively on atmospheric CO_2 levels through further damage to the Earth's capacity for primary productivity.

For terrestrial plant life, the atmosphere is carbon-limiting, and interestingly, one consequence of anthropogenically increased atmospheric CO_2 concentrations is increased agricultural productivity [11, 12]. A return to pre-Industrial Revolution atmospheric CO_2 concentrations of ~ 270–290 ppm [13, 14] would not be welcome from a food production point of view. Currently, the scientific community is recommending a return to 350 ppm [15, 16], from the current levels of approaching 420 ppm [17], and still rising rapidly [13, 18].

To some extent, it is reassuring that the Earth's carbon cycle has naturally compensated for some of our anthropogenic supercharging of the combustion side of the cycle, possibly at the level of 2–3 Gt increased photosynthetic productivity, i.e., around 20% of the anthropogenic imbalance, but it needs our help (**Figure 2**). By building ECOPIA™, we believe we can save the Earth as we want it, and not as we don't want it [19].

2.2 A calculated approach

It is common to think of the sub-polar open oceans with their "boom and bust" annual cycle when characterizing oceanic primary production at the large scale [20, 21]. Cooling and stirring in the winter lead to the replenishment of the surface ocean with essential nutrients as the upper water column is mixed with deeper, nutrient-replete waters. Longer spring sunlight hours re-stratify the upper ocean's thermal structure, making it suitable for photosynthetic phytoplankton growth [22], and the spring phytoplankton bloom begins. Over a period of typically a few days, in any one place, the upper layer nutrients become exhausted, and apart from the mesoscale patchiness [23, 24] created by storm events, frontal boundaries, and eddies, the major primary productivity will await the following winter overturning and a new spring season.

In contrast, however, the large sub-tropical open oceans are dominated by the anticyclonic wind-driven gyres of the low to mid-latitudes [25]. These circulations are convergent in the upper ocean and drive permanent deep mixed layers (**Figure 3**). The permanent mixed layers are so deep (150–300 m) that only rare extreme storm events can penetrate to the high nutrient concentrations of the deep ocean waters; therefore, the surface ocean remains an oligotrophic (nutrient-exhausted) "desert" all year round [20, 26], albeit still supporting diverse microbial communities of comparatively small cell size and abundance [27, 28]. Furthermore, these enormous deserts, one-seventh (~50 million km^2) of the whole of the Earth's ocean area (~362 million km^2), are getting bigger, with high-chlorophyll surface waters being replaced by the increase in the surface areas of the oligotrophic gyres at a rate of 0.8 million km^2 per year [29, 30].

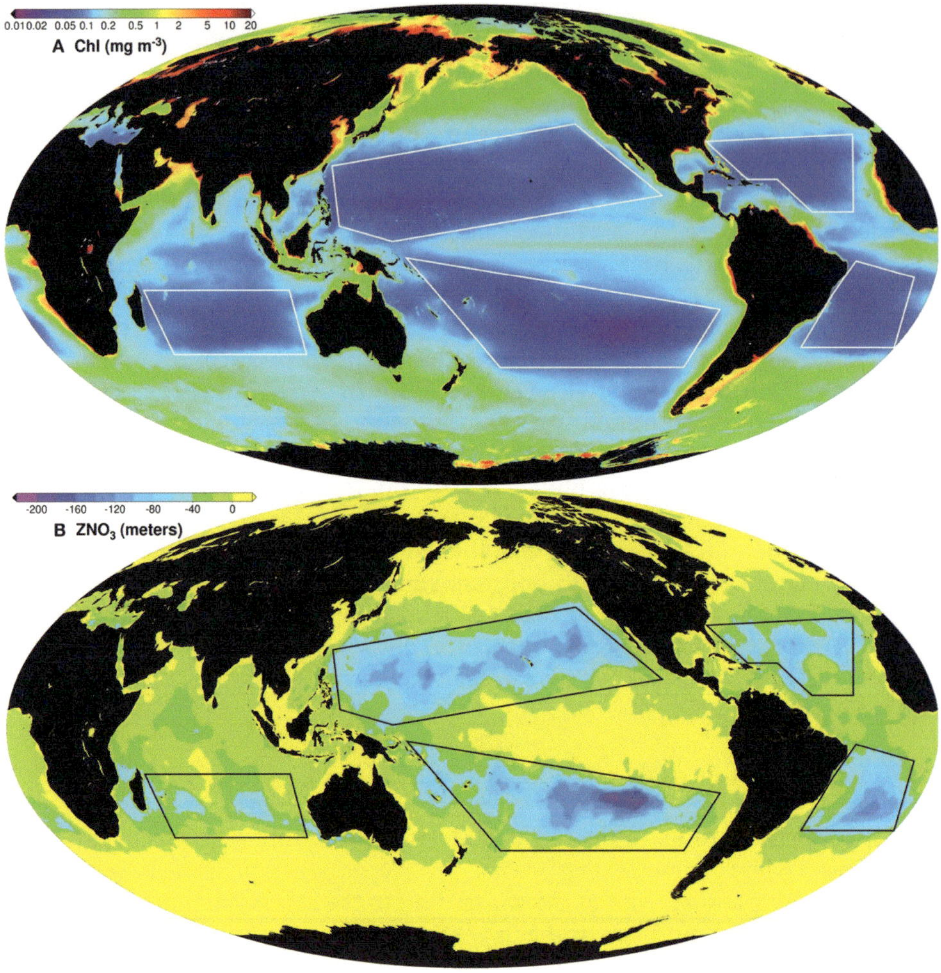

Figure 3.
Global maps of 9 km MODIS a OCI chl-a mission composite (A) and the depth of the 0.2 µm (ZNO_3) nitrate concentration (B) derived from the World Ocean Atlas monthly climatology. Reproduced from Frontiers [30].

The ECOPIA™ program in total, only requires ~0.2 million km^2 of these sub-tropical oligotrophic gyres, just one quarter of their current annual increase in area, to account for an extra ~10 Gigatonnes of carbon per year through photosynthesis: but what is this program? How do we arrive at this number?

ECOPIA™ is made up of Earth Climate Optimization Productivity Island Nodes, henceforth "ecopins", which are modular floating geostationary structures (**Figure 4**). These buoyant structures support light-harvesting mechanisms at the surface, long fiber optic light guides, and a light-distributing mechanism at depth. Each complete ecopin is designed to provide light at depth sufficient to support an annual phytoplankton growth of ~1 million tonnes of chlorophyll, equivalent to a CO_2 uptake of ~100 million tonnes of carbon mass per annum. It is expected that global population growth should level off sometime around 2050, perhaps at 1–2 billion greater than now [3]. Therefore, if we take the, perhaps pessimistic, view that all that can be achieved by the world's population is a static anthropogenic fossil

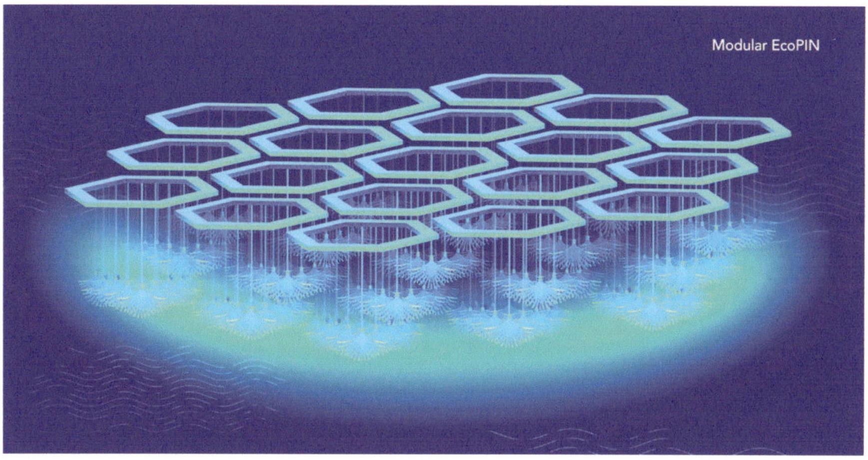

Figure 4.
An artist's impression of a modular ECOPIN.

fuels usage by the year 2030, then we will require around 100 ecopins in total. Of course it is a scalable system, such that if greater reductions in fossil fuels usage can be achieved, then the size of ECOPIA™ reduces approximately linearly.

The detailed engineering design of the ecopins will be determined by marine and civil architects; the light-collecting, transmitting, and diffusing apparatus will be determined by leading glass manufacturing companies working with the architects; and the construction materials will be determined ultimately by expert companies in the field of novel low-carbon-footprint materials science, in collaboration with ship design and marine engineers. Nevertheless, we present our current generic expectations below as a guide.

We propose that each ecopin will be made up of 19 component hexagons (**Figure 4**). Remembering that a regular hexagon is made up of six equilateral triangles, whose base is each a side of the hexagon; then we propose that the central light-collecting "moonpool" will have a hexagonal side dimension of 6,367 m. This will be surrounded by a 500 m wide peripheral platform, such that each component hexagon has a surface area of ~124,400,000 m^2. The component light-collecting area will be ~85% of this, or 105,325,000 m^2, allowing for a slim intra-lens and light accumulator framework. On this basis, each total ecopin (19 component hexagons) will have a surface area of 2,363,600,000 m^2, or 236,360 km^2 for the program (100 ecopins) as a whole, with a light collection area of 200,118 km^2.

It is envisaged that Fresnel lenses will be used in both light collection at the surface and diffusion at depth. These will be focused on sheathed fiber optic bundles. The glass in the lenses, optical coupling, and fiber optic light guides will be a high OH silica glass so that we obtain minimum attenuation in the 350–700 nm visible light spectrum; an example of the commercial availability can be found at [33], but this is just one of many. The fiber optic light guides will need to be 150–300 m long, or more, depending on geographical location. These lengths are such as getting well below the euphotic zone (sunlight-illuminated) and both the biologically active layers of the upper water column [28] (**Figure 5**) and into the high nutrient concentrations of the deeper ocean across the nutricline and permanent thermocline.

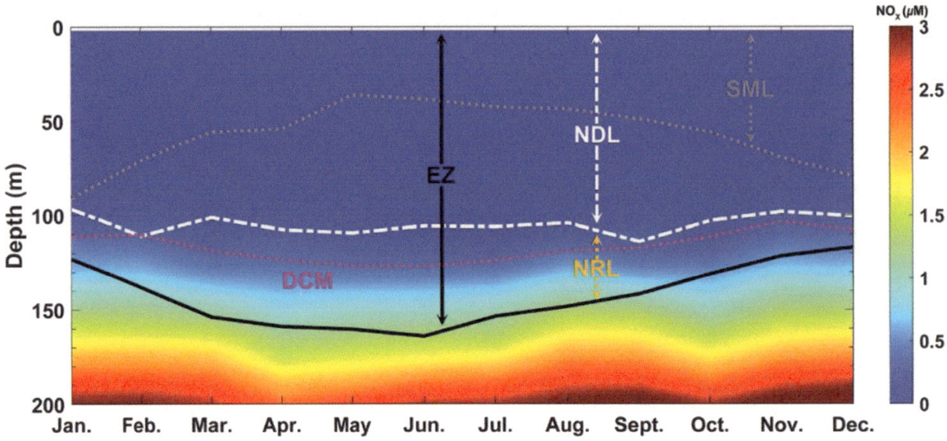

Figure 5.
Reproduced from [28]. Depth of the surface mixed layer (SML), nutrient-depleted layer (NDL), deep chlorophyll maximum, nutrient-replete layer (NRL), and euphotic zone (EZ) at Station ALOHA in the North Pacific Subtropical Gyre. The depth of the NDL is defined as where the nitrate/nitrite (NOx) concentration equals 0.1 μmol L^{-1} and the depth of the EZ is calculated using 1% of the surface usable solar radiation. The background shows the monthly climatology of NOx concentrations. Density, chlorophyll-a, and nutrient data are from the Hawaii Ocean Time Series [31]. EZ data are from [32].

Although ecopins are large structures for our engineers to design and build, they will appear as no more than pinpricks in ocean-color satellite images (**Figure 6**), and they can be built with sustainable materials, drawing down significant further atmospheric CO_2 during construction. It is currently envisaged that the floating peripheral structure supporting and surrounding the light-collecting 'moonpool's will be made of an "eco" friendly material known as "hemp-crete" [34]. Derived from growing hemp on an industrial scale, the construction of ecopins can be made to be a net CO_2 sink procedure [35–37]. Following our paragraphs above, the peripheral structural area in ECOPIA™ is around 36,242 km^2, or 362,420,000 m^2 per ecopin. Assuming 30 m thickness and a 20% "hemp-crete" to air-space ratio, for access, services, storage, etc., this requires a volume of over 2 billion m^3 of "hemp-crete" per ecopin. Even a fairly dense "hemp-crete," say 125 kg m^{-3}, will result in a very buoyant structure.

A full life-cycle assessment (LCA) of hempcrete at the ecopin scale will be essential, although this is beyond the subject of this current manuscript. However, existing data allow us to outline the likely magnitude of the carbon benefit. Hempcrete stores carbon through two mechanisms: biogenic sequestration (0.44–0.55 t C as CO_2 per tonne of dry biomass) in the hemp hurd and mineral carbonation of the lime binder (30–45 kg C as CO_2 per m^3 over the curing period, although it continues to carbonate slowly over decades). This corresponds to a gross biogenic and mineral storage in the order of 60–90 million tonnes C as CO_2 per ecopin, before accounting for transportation, processing, and waterproofing. Indeed, even protecting the hempcrete from the aqueous marine environment could be carried out using a coating of hempseed oil, a plant-derived drying oil which polymerizes on exposure to air and forms a hydrophobic barrier that reduces water ingress without introducing fossil-based polymers.

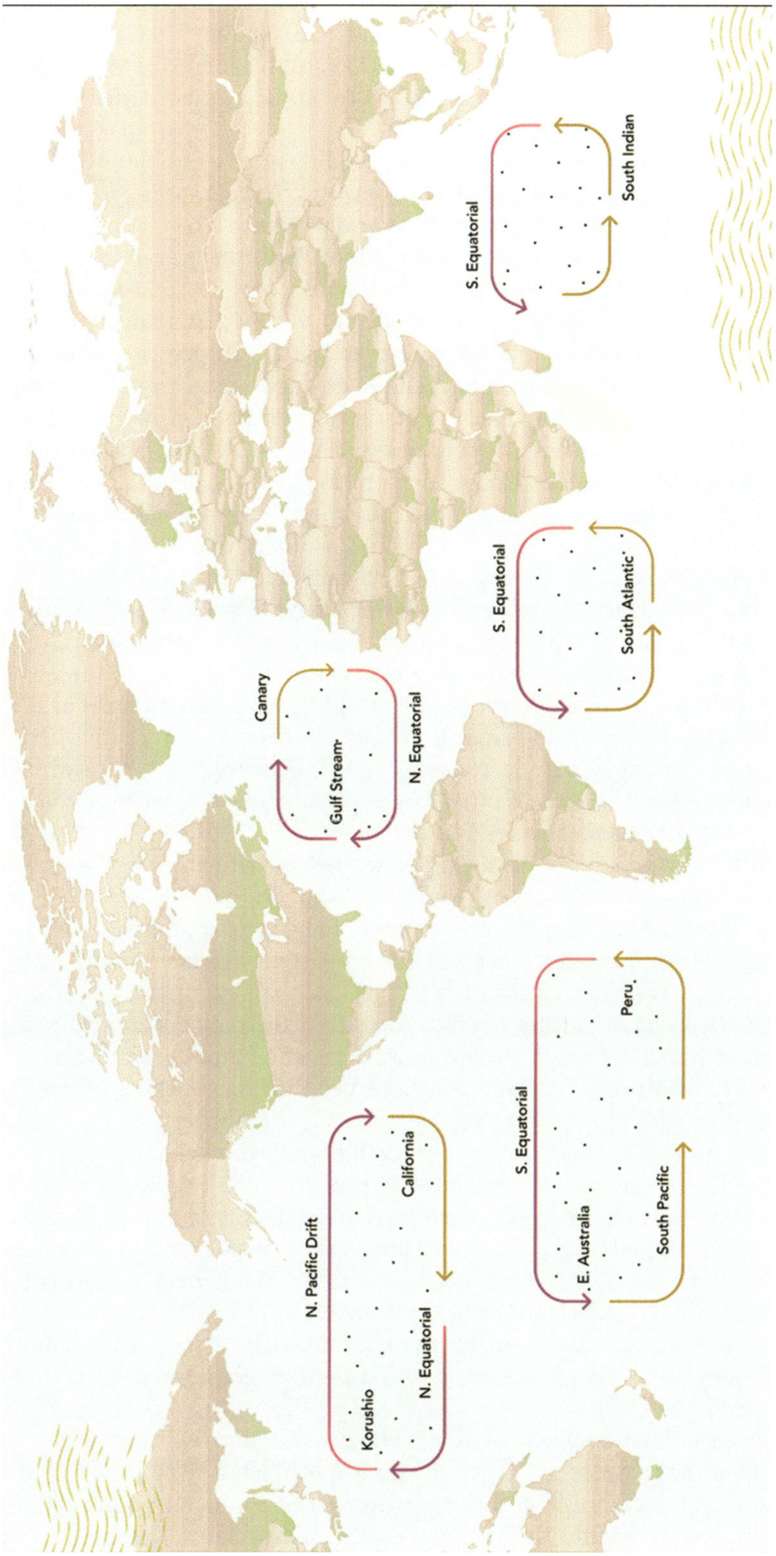

Figure 6.
Ecopins taking up a little over 200,000 km^2: please note this is simply an artist's impression, and nothing should be inferred from the geographical positions of the black "pin-pricks," ecopins, in the cartooned gyres.

More general information about the ECOPIA™ program is available online [38, 39], where both a PDF presentation link and an explainer video are provided, as well as in both published UK patents (GB2601135 and GB2634415) and a published USA patent (No. 18/253574). But how do we get to our large-scale carbon budget and the necessary scale of the program?

Ocean Color data from Earth Observation satellites, for example [40] and **Figure 3**, are useful for assessing the relative contrast and comparison between high and low overall productivity regions of the world's oceans. They clearly show the sub-polar regions as highly productive, with chlorophyll concentrations of around 1 mg m^{-3} as a seasonal or annual mean. However, the temporal and spatial averaging belies the true level of productivity at any one point or time. This can only be reliably measured in-situ. The reason for this is that even in the sub-polar regions, the supplies of nutrients to the surface ocean, either through winter mixing or mesoscale (eddy or frontal) upwelling, are exhausted by photosynthetic growth in very short timescales (days) and at a biological patchiness spatial scale ($\ll$1–10 km) [41–44].

The low-productivity regions of the subtropical gyres are defined by their upper ocean oligotrophic nature and typically have an annual average chlorophyll(a) concentration $\leq$0.1 mg m^{-3} [28, 45]. The euphotic zone of the subtropical gyres is considered to be divided into two layers [46]: a nutrient-depleted layer (NDL), extending from the surface to the top of the nutricline (**Figure 5**), and a nutrient-replete layer (NRL), extending into the upper regions of the nutricline to the base of the EZ [28]. The photosynthetic phytoplankton communities in both these layers tend to be dominated by picoeukaryotic species [47], with growth rates of 0.3–0.9 day^{-1} and ratios of export to primary production (e-ratio) typically <10% compared to >30% in high-latitude oceans [28, 48]. However, in the lower NRL, sporadic phytoplankton blooms of other larger phytoplankton groups, such as diatoms, may enhance these e-ratios [49].

The ECOPIA™ program will take light down below both these surface layers, across the nutricline, to create a new "euphotic" zone in the deep ocean, of, in practice, limitless nutrient supply, that is slowly moving past geostationary light sources, ecopins; thus, providing an ideal and consistent diurnal growth environment. Here, we expect very high e-ratios, as the new primary production is capped below the permanent thermocline and out of the range of mixing processes with the surface ocean. Of course, these e-ratios must be monitored under a comprehensive Measurement, Reporting, and Verification (MRV) protocol [50], in this case, ECOPIA's MRV protocol [51].

In-situ measurements of sub-polar latitude blooms, whether following over-winter mixing [52] or mesoscale eddy stirring processes [53], typically result in blooms of 5–12 mg m^{-3} chlorophyll concentration, as the nutrient supply is still only sufficient rather than plentiful [54]. A deep chlorophyll maximum may frequently be observed just above the thermocline if it is still within the light-penetrating photic zone, and chlorophyll concentrations of typically up to 1 mg m^{-3} may be maintained here for some time, again due to sufficient vertical mixing across the thermocline, but also low-light adaptation processes within the phytoplankton cellular photo mechanisms [55].

Mesocosm experiments, where artificial water columns are seeded with phytoplankton and the necessary nutrients for growth, have been shown to produce phytoplankton bloom chlorophyll concentrations of 60 mg m^{-3} or more [56] when a plentiful supply of nutrients is provided. Furthermore, a day^{-1} doubling rate can be achieved under these conditions even with modest light supplies of 200 μmole photons s^{-1} m^{-2}, which corresponds to the typical maximum irradiance phytoplankton are acclimatized to accept in the turbulent surface mixed layer before becoming photo-saturated [57].

Daylight surface light levels available for capture at subtropical latitudes are typically 1,500–3,000 μmole photons s^{-1} m^{-2} [58]. Allowing for an expected 50% propagation loss over the (150–300 m) light fiber bundles, Ecopins will supply 750–1500 μmole photons s^{-1} m^{-2} and even greater losses would still provide significantly more than the photo-saturation level of 200 μmole photons s^{-1} m^{-2}, discussed above. This light will be delivered by a photo-diffuser over a depth range of ~20 m or greater and an assumed worst-case photo propagation distance, euphotic depth [59, 60], of 30 m. If we conservatively assume just half of the bloom chlorophyll concentrations from mesocosm experiments discussed above, i.e., 30 mg m^{-3}, then, over a 50 m "photic" depth at a day^{-1} doubling rate, this equates to 1.5 g m^{-2} day^{-1} chlorophyll growth, or >500 g m^{-2} $year^{-1}$. Demonstrator programs discussed in section 4 will be used to test the sensitivity of light supply and biological response.

Nutrients remain plentiful as a result of the geostationary design of the ecopins and the natural currents flowing past at depth, thus preventing any noticeable decrease in nutrient concentrations over time. Ocean currents, even 100–300 m below the surface of the ocean, are typically around 5–20 cm s^{-1} and, therefore, move water distances of anywhere from 5 to 20 km per day: over a 50 km diameter ecopin, even the slowest of these currents would only provide light for a 10-day bloom period.

In any location, carbon to chlorophyll (C:Chl) ratios in marine phytoplankton are widely variable, depending on species, light levels, nutrient availability, seasonality, and combinations of these and other factors [61, 62]. Published ratios of >10 to <300 have been carefully calculated through a combination of observation and modeling. Assuming the provision of light by ecopins initiates transient bloom phases of dominantly small-cell plankton species [47] in the waters below the traditional photic zone of the oligotrophic gyres, a modest C:Chl ratio of 100, typically 80–150 reported for these species [61], gives us a carbon uptake of 50 kg C m^{-2} $year^{-1}$. This equates to 98 megatonnes of carbon per ecopin per year, or 9.8 Gigatonnes per year by the ECOPIA™ program in total (100 ecopins). However, the C:Chl ratio is clearly a very sensitive part of this calculation, and monitoring this through sampling and active fluorescence observations will be an important component of the ECOPIA™ MRV protocol [51]. We have not yet considered low-light adaptation of the phytoplankton photo mechanisms, mentioned earlier [55], which may act to significantly increase the volume of the effective photic zone created by each ecopin, precisely because this additional productivity may come at a cost of a lower C:Chl ratio [62].

Approaching our calculation from the other direction, it is generally considered that approximately 10% of surface ocean primary production is exported below the thermocline [63] as detritus and fecal pellets. The ECOPIA™ program transports light below the thermocline to allow new production to take place below the thermocline, where the resulting particulate organic carbon (POC), consisting of detritus and fecal pellets, is by traditional definition already exported. That way, it initially makes sense to consider that the ecopins might only require 10% of the area they would require at the surface if nutrients were available. Furthermore, the productive surface ocean is generally only nutrient-sufficient, and only for the typically short duration of a bloom, as discussed earlier. Whereas below the thermocline, nutrients are plentiful, and water is continuously flowing past, so the ECOPIA™ program provides a continuous growth season. Thus, it is not unreasonable to conclude that less than 0.06% of the total ocean surface (~0.40% of the surface area of the oligotrophic gyres) is required to create the 10% increase in oceanic carbon export that is needed to re-balance the Earth's carbon cycle, as estimated in the forward argument above.

2.3 A diligent approach

The Earth's large-scale carbon cycle is well known; however, many of the nature-based solutions historically considered reside in the details of the cycle, and these are not always so well known. For just two examples, the concept of pumping deep ocean waters to the surface looked promising; however, it carried the problem that the process can't help but pump the sinking sequestered export carbon (organic carbon) back up to the surface at the same time [64, 65]. Fertilizing the circumpolar Southern Ocean with the micronutrient, iron, was considered for several years, until it generally became recognized that this might not support biodiversity in planktonic species [66–68]. These uncertainties largely arose from the "farming"-like practice of changing the composition of the "soil," or in this case, the ocean waters. The ECOPIA™ program takes a different approach, bringing light down to the depths where there are plenty of naturally determined nutrients. This program simply provides light, and nothing else, to a naturally diverse seed population.

There are still uncertainties, of course; how effectively will the program support secondary production and the creation of new fisheries? Is there a significant lag timescale in drawing atmospheric CO_2 out of the atmosphere? And what happens to the complex processes of export production, i.e., particle transformation, remineralization, microbial respiration, etc., now these are moved below the thermocline? Critically, the scale of the ecopin components makes it possible to monitor and to model these uncertainties. As discussed earlier, ocean currents, even 100–300 m below the surface of the ocean, move water distances of anywhere from 5 to perhaps as much as 20 km per day; the individual geostationary ecopins are chosen to be the size described earlier, so that measurements of photosynthetic growth, biodiversity, and CO_2 uptake can be made over a growth time scale of at least 24–48 hours and up to perhaps as much as 10 days.

Three-dimensional advective processes and turbulence are ubiquitous in the oceans; these are the equivalent of stirring in a cup of coffee and speed up the effective mixing timescale by as much as two orders of magnitude compared to that of pure diffusivity. Recent quantification estimates of coherent three-dimensional advection processes [69, 70] indicate that they could account for a significant proportion of the export of organic carbon to the deep ocean, alongside gravitational sinking and turbulent mixing.

As early as the mid-1980s, we knew that anthropogenic CO_2 had made its way throughout the entire depth of the North Atlantic Ocean oligotrophic subtropical gyre [71]. From mid-1990s data [72], anthropogenic CO_2 had quite clearly reached a depth of over 1500 m in the Indian Ocean oligotrophic subtropical gyre (**Figure** 7).

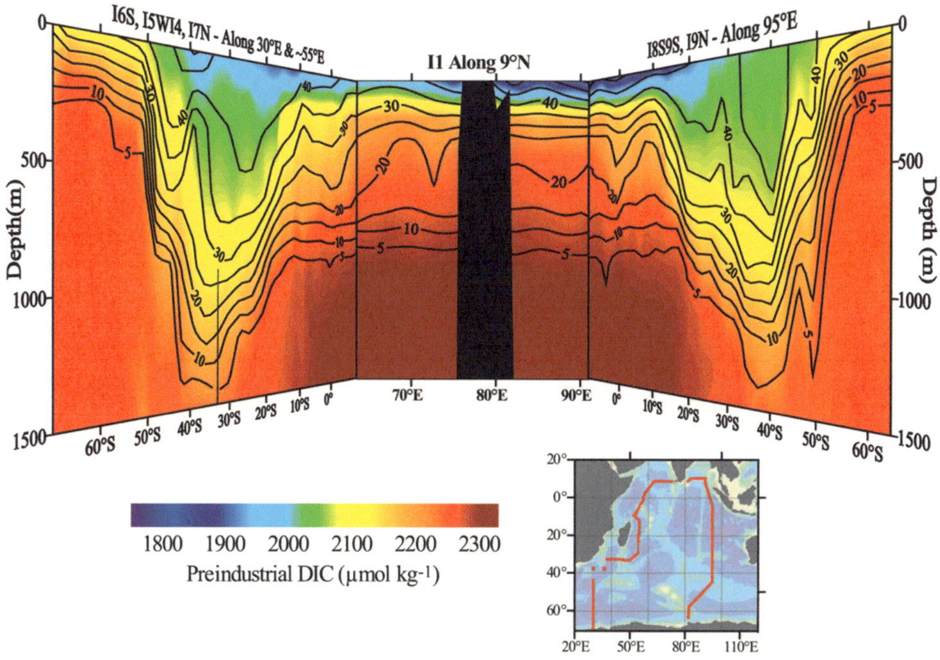

Figure 7.
Contoured sections of preindustrial CO_2 (colors) and anthropogenic CO_2 (contours) – reproduced from [72].

Assuming an exponential growth in anthropogenic CO_2 production since the mid-1880s [73], these studies provide evidence for an equivalent vertical diffusion timescale of at least on the order of 100 m year^{-1}. Even in the most conservative of the oligotrophic ocean gyres, those in the Pacific, where the anthropogenic input is perhaps at the greatest distance from its source, the difference between the 1990s WOCE experiment data and the 2000s CLIVAR experiment data for delta13 ($^{13}C/^{12}C$) CO_2 concentrations [74] indicated an anthropogenic CO_2 uptake rate of almost a billion tonnes per year in total, with the largest proportion of this at the subtropical oligotrophic latitudes. The ECOPIA™ program will work alongside and assist this natural biological restoring pump, which is clearly capable of handling the scale of the challenge.

Early unpublished modeling of light guide tele-illumination used an existing medium-resolution physical ocean model, NEMO [75], at 1-degree resolution. An existing biogeochemical model with a carbon cycle, MEDUSA-2.0 [76], coupled to this, showed significant increases in CO_2 fluxes across the surface ocean from the atmosphere with small quantities of downwelling light (10%) from the ocean's surface to the base of the euphotic layer (**Figure 8**). This modeled tele-illumination enhanced ocean primary production, export production, and the uptake of anthropogenic CO_2; it also decreased ocean acidification. However, whilst total primary production was enhanced, near-surface production was commensurately reduced (not shown) by the transfer of light to depth, which is why the oligotrophic gyres, where surface production is limited by a lack of nutrients, are the ideal place to position ecopins. Naturally, initial and operational modeling programs are essential components of ECOPIA™ as discussed in Section 4 and the MRV protocol [51].

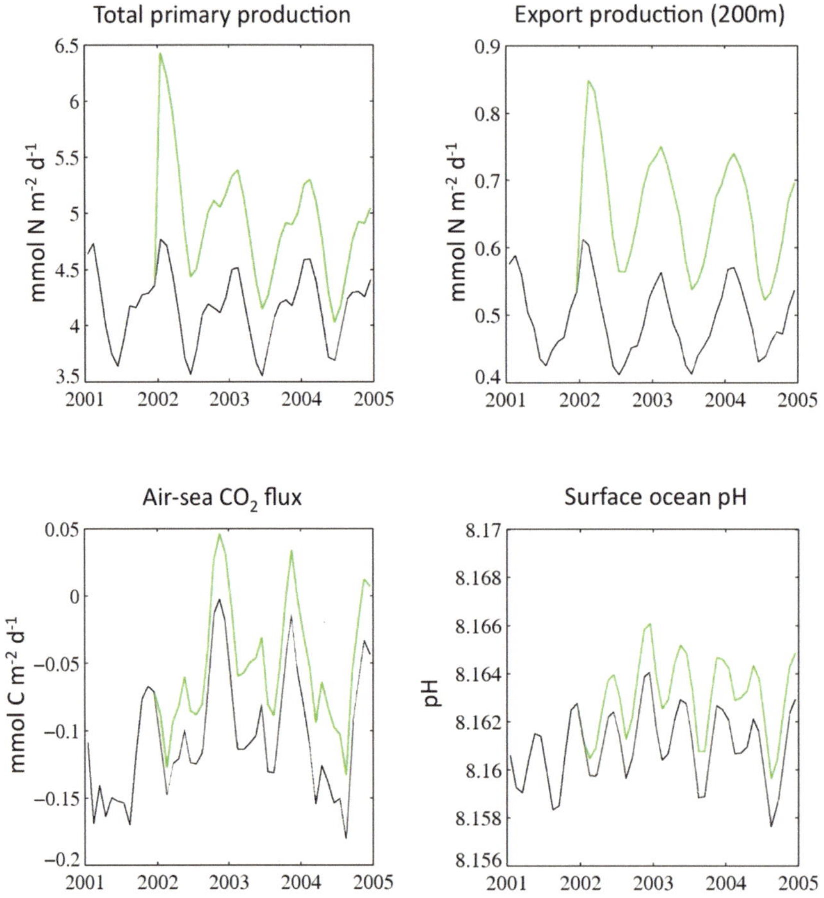

Figure 8.
Early modeling results (globally averaged) for total primary production, export production through 200 m, air-sea CO_2 flux into the ocean, and surface ocean pH, where the control is given by the black line, and 10% light transferred to the base of the euphotic zone is given by the green line (Andrew Yool, 2022, personal communication).

3. Economic case

A report by [77] stated that capital spending on physical assets for energy and land-use systems in the net-zero transition between 2021 and 2050 would amount to about $187 trillion to decarbonize 100% of the global economy. In addition, only the Global North can afford to decarbonize their economies. Data from "Our World in Data" [78] shows that those "developing" nations that invest the most heavily in energy production show the greatest increase in GDP per capita. A standout example is China, which added an additional 19,000 kWh per capita in terms of energy consumption between 1995 and 2020 and saw an increase of $15,211 per capita (US Dollar at 2017 value) over the same period, a relative increase of 636%. This creates a large amount of pressure against net-zero decarbonization efforts, as the cheapest and quickest ways to create additional energy baseload production are often via "dirty" means, such as fossil fuels [79, 80]. Clearly, the levels of projected capital spending required, and the need for "developing" nations to improve the lives of their citizens, make attempts to achieve net-zero globally through decarbonization alone seem unrealistic and unachievable at present. The situation makes the case that we need to learn to live with and manage our carbon output as a global society, thereby mitigating the climate crisis.

It is critical to stop the externalization of the costs of CO_2 emissions by industry and society as a whole. As an example of "offshoring," as it has been termed by some, the UK's territorial emissions carbon footprint decreased from 218.16 Mt of carbon in 1990 to 115.90 Mt in 2021; however, conversely, the UK's consumption-based carbon footprint increased from 231.80 Mt in 1990 to 291.79 Mt of carbon in a similar period [81–83], and this is just a single, not the worst, example. This will necessitate the implementation of regulation requiring that companies (and, in the end, consumers) pay the costs of CO_2 emissions through the internalization of these costs back onto the company balance sheets, most likely via a form of carbon tax and/or carbon tariff. Given the cost and complexities of complete decarbonization there will need to be some mechanism that allows companies to obtain carbon offset credits to offset their remaining carbon footprint and thereby mitigate the costs of the carbon tax or tariffs. Carbon emissions of companies will need to be monitored, and enforcement put in place, for companies to ensure they have purchased the required amount of credits and that the credits they have purchased are from truly additive nature-based solutions. A coordinated international effort will ensure that multinationals cannot dodge their obligations and that there is a global response to a global crisis.

A hybrid public-private financing model is critical to solve the global climate crisis. We believe this should be built around a regulated carbon offset credit market. In particular, governments should create demand for carbon offset credits via mechanisms such as carbon taxes and tariffs, and should assist in infrastructure financing (to help build effective nature-based solutions) through mechanisms such as bonds backed by the funds raised through these carbon taxes and tariffs. The cost of a carbon offset credit is projected to be between £290 and £550 (GBP) per tonne of carbon by 2050, according to the UK Government [84]. In February 2023, the market for carbon permits in the European Union rose above €385 ($415) per tonne of carbon for the first time [85]. Calculations of the cost of constructing a full-scale implementation of the ECOPIA™ program put the figure at circa $10 trillion, much of which would be covered by carbon offset credits. The model will work even at a price of just $50 per tonne. A large sum of money but small compared to the projected cost of a real net-zero transition discussed earlier [77]. Once constructed, this program and other private enterprises should be self-financing from operations trading through carbon offset credits via regulated exchanges.

Marine Carbon Dioxide Removal (mCDR) has alternatives, including Ocean Enhanced Alkalinity (OAE) and Direct Ocean Capture (DOC), alongside the nature-based pathway of ECOPIA™. OAE sequesters bicarbonate and carbonate ions deeper in the ocean, either through adding alkaline chemicals or electrolysis. Initial cost estimates range from ~$990 to $1,290 per tonne C as CO_2 with projections of at-scale costs of ~$185–$590 per tonne C as CO_2. DOC strips dissolved CO_2 from seawater and sequesters it in some form of storage. Initial cost estimates range from ~$1,600 to $6,750 per tonne C as CO_2 with projections of at-scale costs of ~$370 to $740 per tonne C as CO_2 before storage costs. In comparison, ECOPIA™ initial cost estimates are similar to those of DOC's current costs, ~$1500 to $5,000 per tonne C as CO_2, with projections of at-scale costs of $50–$150 per tonne C as CO_2, below those of OAE at-scale costs.

Finally, perhaps it is instructive to recall the 80:20 rule, or Pareto's principle, which observes that, generically, around 80% of consequences come from typically around 20% of the potential causes [86, 87]; often also thought of as a component of the principle of diminishing returns [88]. Under this concept, we expect that most of net zero can be achieved for circa $20 trillion [77], and incorporating ECOPIA™ at the required scale would bring this to approximately one-seventh of the cost of global net zero. ECOPIA™ is designed to solve the anthropogenic climate change problem even if the best the global community can achieve regarding emissions reductions is a static fossil fuels usage by the year 2030, as stated in section (2.2). If, on the other hand, a high percentage of net zero can really be achieved, then the required number of the component ecopins will be reduced and/or the legacy atmospheric CO_2 concentration can be reduced more quickly.

4. Conclusion

The ECOPIA™ program is a major engineering program, but not a geo-engineering program; it is a Nature-Based Carbon Cycle Management Solution (NBCCMS). It makes no attempt to change the chemistry of the ocean so that it can sustainably support a faster solubility pump; quite the reverse, the program naturally supports the biological pump to return the solubility pump back to carbon cycle balance. Indeed, when completed, the final engineering product will be an adjustable Earth "climostat" to maintain the atmospheric CO_2 concentration where the global society requires it to be. The program is an NBCCMS for dealing with the atmospheric anthropogenic CO_2 overload, the primary waste product of our industrial creativity. It does not, in itself, capture CO_2 neither does it stockpile CO_2 in high-pressure reservoirs or complex chemical equilibria. The program simply provides light at the right depth, in a small, scattered proportion of the world's oceans' largest deserts, to allow natural biodiverse community uptake of dissolved CO_2 through photosynthesis, the natural process that completes the carbon cycle. By thus allowing the ocean to naturally increase its photosynthetic capacity by 10%, using less than 0.002% of its volume, the distributed ecopins will enable the oceans to restore the Earth's carbon cycle balance themselves.

The final export (sequestration) of the carbon will be to the deep sea, and eventually, for some, the deep seabed from whence it came, as a natural deposition with a lifetime of hundreds to millions of years. The capacity of the deep sea is large, with a current store of approximately 37,000 billion tonnes of carbon [1]. There have been three important recent papers [89–91], which, in different ways, show how important the ECOPIA™ program is to actively support the re-balancing of the Earth's carbon cycle. The first shows that, left alone, the ocean will carry on its slow adjustment until around 2100, but then it will stop and could even become a source of carbon, as the combination of temperatures, acidification, and climate change-driven diversity limitation has a significantly negative effect. The details lie in the changes to the twilight zone, 100–1,000 m depth, which is still currently relatively poorly understood and modeled. The second and third papers show that the export of synthesized carbon (organic carbon resulting from photosynthesis) below 1000 m is critical for long-term storage over 100 years, and hence the depth of enhanced photosynthesis achieved by the distributed ecopins is so critical. Hence, the importance of ECOPIA's Measurement, Reporting, and Verification (MRV) protocol [51], as mentioned earlier.

Continued and accelerated research into how the ocean will continue to naturally absorb carbon dioxide through photosynthesis is essential. Demonstrator and modeling programs will begin in 2026 and 2027 and will continue to run in parallel with the initial construction of the ECOPIA™ program. The modeling program incorporates at least two different model structures to differentiate between inherent model biases and the transport and mixing processes being modeled. The demonstrator program involves the construction of scaled-down ecopins, testing light level provision, induced primary productivity, biodiversity, and export.

The ECOPIA™ program has identified five research programs that will run in parallel with the construction of the Climostat. These are:

1) Multi-disciplinary (biogeochemistry and physics) coupled ocean/atmosphere system modeling, involving digital triplets of the forecast results of inaction, action of building ecopins, and assimilated reality.

2) Researching the optimization of optical collectors, transmitters, and diffusers for light in the visible spectrum.

3) Investigating hempcrete and other sustainable plant-based alternatives for maritime construction materials science.

4) Updating the inventory of marine Delta 13 Carbon measurements to support hindcast and forecast modeling of the spread of anthropogenic CO_2 through the deep ocean.

5) Investigating export processes through the mesopelagic ocean. This study is designed to contribute to the scientific understanding of the complex processes of export production, including particle transformation, remineralization and microbial respiration.

Calculations around the cost to construct a full-scale implementation of ECOPIA™ put the figure for construction at circa $10 trillion, a fraction of the cost to implement Net-Zero as laid out by [77]. Given that the projected cost of a Climostat such as ECOPIA™, which would allow us to live with and manage our carbon output for an implementation that would cater for the projected global emissions at 2030 levels, is so comparatively "reasonable", it would seem to make economic sense to look to provision such a system.

Ecopins will allow the Earth's oceans to breathe just a little more clearly, reducing the light restriction deep beneath just 236,360 km^2 of some of the otherwise most unproductive regions. This program maintains and encourages natural biodiversity to achieve this, and furthermore, natural nutrient balances are left untouched. The ECOPIA™ program simply provides light. MyOcean Resources Ltd. is convinced that since society worked out how to boost the combustion side of the Earth's carbon cycle, it doesn't seem too farfetched to imagine that society can implement NBCCMSs for boosting the photosynthetic side of this natural cycle and rebalancing the system.

Acknowledgments

We are most grateful for informal reviews and discussions of our program by friends and colleagues, particularly for their helpful comments, suggestions, and feedback. And we are indebted to the reviewers, whose comments have been most insightful.

Conflict of Interest

The authors declare that this research review was conducted in the absence of any external commercial or financial relationships. Internally, the authors are, of course, connected with the Ecopia Marine Ltd. program at either directorial or advisory level through MyOcean Resources Ltd., as clearly declared in the author affiliations.

Author contribution

Sections 1, 2, and 4 were researched and written by Allen, but with substantial input and comments from Fitzgerald and Franks. Section 3 was researched and written by Fitzgerald, with substantial input and comments from Franks and some input and comments from Allen.

Author details

John Allen*, Calum Fitzgerald and Lonnie Franks
My Ocean Resources Limited, Hornchurch, UK

*Address all correspondence to: john@myocean.co.uk

References

[1] UCAR Center for Science Education Carbon cycle diagram from the DOE with numbers [Internet]. 2025. Available from: https://scied.ucar.edu/image/carbon-cycle-diagram-doe-numbers [Accessed: 2025-April-28]

[2] Post WM, Peng T-H, Emanuel WR, King AW, Dale VH, Deangelis DL. The global carbon cycle. American Scientist. 1990;**78**:310–326. DOI: 10.1007/978-3-642-84608-3_10

[3] Lutz W, Samir K. Dimensions of global population projections: What do we know about future population trends and structures? Philosophical transactions of the royal society of London. Philosophical Transactions of the Royal Society B: Biological Sciences. 2010;**365**:2779–2791. DOI: 10.1098/rstb.2010.0133

[4] Watson AJ. Certainty and uncertainty in climate change predictions: What use are climate models? Environmental and Resource Economics. 2008;**39**(1):37–44. DOI: 10.1007/s10640-007-9173-9

[5] Goldblatt C, Watson AJ. The runaway greenhouse: Implications for future climate change, geoengineering and planetary atmospheres. Philosophical Transactions of the Royal Society A: Mathematical, Physical and Engineering Sciences. 2012;**370** (1974):4197–4216. DOI: 10.1098/rsta.2012.0004

[6] Riding R. The term stromatolite: Towards an essential definition. Lethaia. 2007;**32**:321–330. DOI: 10.1111/j.1502-3931.1999.tb00550.x

[7] Riding R, Reitner J, Quéric N-V. The nature of stromatolites: 3,500 million years of history and a century of research. Ed. Arp, Gernot. 2011. DOI: 10.1007/978-3-642-10415-2_3

[8] Popall RM, Bolhuis H, Muyzer G, Sánchez-Román M. Stromatolites as biosignatures of atmospheric oxygenation: Carbonate biomineralization and UV-C resilience in a geitlerinema sp.-Dominated culture. Frontiers in Microbiology. 2020;**11**. DOI: 10.3389/fmicb.2020.00948

[9] IPCC. Climate Change 2022: Impacts, Adaptation, and Vulnerability. Contribution of Working Group II to the Sixth Assessment Report of the Intergovernmental Panel on Climate Change. In: Pörtner H-O, Roberts DC, Tignor M, Poloczanska ES, Mintenbeck K, Alegría A, Craig M, Langsdorf S, Löschke S, Möller V, Okem A, Rama B, eds. Cambridge, UK and New York, USA: Cambridge University Press; 2022. p. 3056. DOI: 10.1017/9781009325844

[10] Barker S, Higgins JA, Elderfield H. The future of the carbon cycle: Review, calcification response, ballast and feedback on atmospheric CO_2. Philosophical Transactions of the Royal Society of London. Series A, Mathematical, Physical and Engineering Sciences. 2003;**361**:1977–1999. DOI: 10.1098/rsta.2003.1238

[11] Kimball BA, Idso SB. Increasing atmospheric CO_2: Effects on crop yield, water use and climate. Agricultural Water Management. 1983;7(1–3):55–72. DOI: 10.1016/0378-3774(83)90075-6

[12] Deryng D, Elliott J, Folberth C, Müller C, Pugh TAM, Boote KJ, Conway D, Ruane AC, Gerten D, Jones JW, Khabarov N, Olin S, Schaphoff S, Schmid E, Yang H, Rosenzweig C. Regional disparities in the beneficial effects of rising CO_2 concentrations on crop water productivity. Nature Climate Change. 2016;**6**(8):786–790. DOI: 10.1038/nclimate2995

[13] World Economic Forum, Nature and Biodiversity. Met Office: Atmospheric CO_2 now hitting 50% higher than pre-industrial levels. 2021. Available from: https://www.weforum.org/agenda/2021/03/met-office-atmospheric-co2-industrial-levels-environment-climate-change/ [Accessed: 2025-April-29]

[14] Ciais P, Sabine C, Bala G, Bopp L, Brovkin V, Canadell J, Chhabra A, DeFries R, Galloway J, Heimann M, Jones C, Le Quéré C, Myneni RB, Piao S, Thornton P. Carbon and Other Biogeochemical Cycles. Chapter 6 In: Climate Change 2013: The Physical Science Basis. Contribution of Working Group I to the Fifth Assessment Report of the Intergovernmental Panel on Climate Change. In: Stocker TF, Qin D, Plattner G-K, Tignor M, Allen SK, Boschung J, Nauels A, Xia Y, Bex V, Midgley PM, eds. Cambridge, United Kingdom and New York, USA: Cambridge University Press; 2013. https://www.ipcc.ch/site/assets/uploads/2018/02/WG1AR5_Chapter06_FINAL.pdf

[15] Hansen J, Sato M, Kharecha P, Beerling D, Berner R, Masson-Delmotte V, Pagani M, Raymo M, Royer DL, Zachos JC. Target atmospheric CO_2: Where should humanity aim? The Open Atmospheric Science Journal. 2008;**2**:217–231. DOI: 10.2174/1874282300802010217

[16] Veron JEN, Hoegh-Guldberg O, Lenton TM, Lough JM, Obura DO, Pearce-Kelly P, Sheppard CRC, Spalding M, Stafford-Smith MG, Rogers AD. The coral reef crisis: The critical importance of < 350ppm CO_2. Marine Pollution Bulletin. 2009;**58**(10):1428–1436. DOI: 10.1016/j.marpolbul.2009.09.009

[17] IPCC. Climate Change 2021: The Physical Science Basis. Contribution of Working Group I to the Sixth Assessment Report of the Intergovernmental Panel on Climate Change. In: Masson-Delmotte V, Zhai P, Pirani A, Connors SL, Péan C, Berger S, Caud N, Chen Y, Goldfarb L, Gomis MI, Huang M, Leitzell K, Lonnoy E, Matthews JBR, Maycock TK, Waterfield T, Yelekçi O, Yu R, Zhou B, eds. Cambridge, United Kingdom and New York, NY, USA: Cambridge University Press; 2021. DOI: 10.1017/9781009157896

[18] OECD. OECD Environmental Outlook to 2050: The Consequences of Inaction. Paris: OECD Publishing; 2012. DOI: 10.1787/9789264122246-en

[19] Allen J, Fitzgerald C, Franks L. Re-balancing the Earth's Natural Carbon Cycle; Greening the Deserts of the Oceans through Tele-illumination. In: EGU General Assembly 2023; Vienna, Austria; 23–28 Apr 2023. p. EGU23–8864. DOI: 10.5194/egusphere-egu23-8864

[20] Longhurst AR Ecological Geography of the Sea. 2nd edn. San Diego, CA: Academic Press; 1988. DOI: 10.1016/B978-0-12-455521-1.X5000-1

[21] Siegel DA, Doney SC, Yoder JA. The North Atlantic spring phytoplankton bloom and Sverdrup's critical depth hypothesis. Science. 2002;**296**:730–733. DOI: 10.1126/science.1069174

[22] Henson SA, Sanders R, Holeton C, Allen JT. Timing of nutrient depletion, diatom dominance and a lower-boundary estimate of export production for Irminger Basin, North Atlantic. Marine Ecology Progress Series. 2006;**313**:73–84. DOI: 10.3354/meps313073

[23] Denman KL, Dower JF, He X. Patch Dynamics. In: Cochran JK, Bokuniewicz HJ, Yager PL. editors. Encyclopedia of Ocean Sciences. London: Third. Academic Press; 2019. p. 765–770. DOI: 10.1016/B978-0-12-409548-9.10805-X

[24] McGillicuddy DJ, Franks PJS. Models of Plankton Patchiness. In: editors, Cochran JK, Bokuniewicz HJ, Yager PL. Encyclopedia of Ocean Sciences, Vol. **2019**. London: Third. Academic Press; 2019. p. 536–546. DOI: 10.1016/B978-0-12-409548-9.11610-0

[25] Stommel H. The westward intensification of wind-driven ocean currents. EOS. 1948;**29**(2):202–206. DOI: 10.1029/TR029i002p00202

[26] Morel A, Claustre H, Gentili B. The most oligotrophic subtropical zones of the global ocean: Similarities and differences in terms of chlorophyll and yellow substance. Biogeosciences. 2010;7:3139–3151. DOI: 10.5194/bg-7-3139-2010

[27] Yin Q, Fu B, Li B, Shi X, Inagaki F, Zhang X-H. Spatial variations in microbial community composition in surface seawater from the ultra-oligotrophic center to rim of the South Pacific Gyre. PLoS ONE. 2013;**8**(2). DOI: 10.1371/journal.pone.0055148

[28] Dai M, Luo Y-W, Achterberg EP, Browning TJ, Cai Y, Cao Z, Chai F, Chen B, Church MJ, Ci D, Du C, Gao K, Guo X, Hu Z, Kao S-J, Laws EA, Lee Z, Lin H, Liu X, Luo W, Meng F, Shang S, Shi D, Saito H, Song L, Wan XS, Wang Y, Wang W-L, Wen Z, Xiu P, Zhang J, Zhang R, Zhou K. Upper ocean biogeochemistry of the oligotrophic North Pacific Subtropical Gyre: From nutrient sources to carbon export. Reviews of Geophysics. 2023;**61**: e2022RG000800. DOI: 10.1029/2022RG000800

[29] Polovina JJ, Howell EA, Abecassis M. Ocean's least productive waters are expanding. Geophysical Research Letters. 2008;**35**(3). DOI: 10.1029/2007GL031745

[30] Signorini S, Franz B, McClain C. Chlorophyll variability in the oligotrophic gyres: Mechanisms, seasonality and trends. Frontiers in Marine Science. 2015;**2**. DOI: 10.3389/fmars.2015.00001

[31] Hawaii Ocean Time-series HOT-DOGS application; University of Hawai'i at Mānoa. National Science Foundation Award # 2241005. 2025. Available from: https://hahana.soest.hawaii.edu/hot/hot-dogs/ [Accessed: 2025-April-29]

[32] Wu B, Lin X, Yu L. Poleward shift of the kuroshio extension front and its impact on the North Pacific subtropical mode water in the recent decades. Journal of Physical Oceanography. 2021;**51**(2):457–474. DOI: 10.1175/JPO-D-20-0088.1

[33] Avantes member of the Nynomic group, Introduction to Fiber Optics.

2025. Available from: www.avantes.com/support/theoretical-background/introduction-to-fiber-optics/ [Accessed: 2025-April-29]

[34] Özdamar EG. Hemp as a potential material in architecture: Is it possible in Turkey? ICONARP International Journal of Architecture and Planning. 2021;**9**(1):131–154. DOI: 10.15320/ICONARP.2021.153

[35] Brar KK, Raheja Y, Chadha BS, Magdouli S, Brar SK, Yang Y-H, Bhatia SK, Koubaa A. A paradigm shift towards production of sustainable bioenergy and advanced products from Cannabis/hemp biomass in Canada. Biomass Conversion and Biorefinery. 2022;**14**. DOI: 10.1007/s13399-022-02570-6

[36] Zatta A, Monti A, Venturi G. Eighty years of studies on industrial hemp in the po valley (1930–2010). Journal of Natural Fibers. 2012;**9**(3):180–196. DOI: 10.1080/15440478.2012.706439

[37] Amaducci S, Zatta A, Pelatti F, Venturi G. Influence of agronomic factors on yield and quality of hemp (Cannabis sativa L.) fibre and implication for an innovative production system. Field Crops Research. 2008;**107**(2):161–169. DOI: 10.1016/j.fcr.2008.02.002

[38] MyOcean Resources. Our Projects – Ecopia. 2021. Available from: https://www.myocean.co.uk/#ourProjects [Accessed: 2025-April-29]

[39] Ecopia Marine. Green Oases in the Deserts of the Ocean. 2022. Available from: https://www.ecopia.world/ [Accessed: 2025-April-29]

[40] NASA Ocean Biology Processing Group. Ocean Color Web. 2023. Available from: https://oceancolor.gsfc.nasa.gov/ [Accessed: 2025-April-29]

[41] Mackas D, Denman K, Abbott M. Plankton patchiness: Biology in the physical vernacular. Bulletin of Marine Science. 1985;**37**:652–674

[42] Martin AP. Phytoplankton patchiness: The role of lateral stirring and mixing. Progress in Oceanography. 2003;**57**(2):125–174. DOI: 10.1016/S0079-6611(03)00085-5

[43] Lévy M, Franks PJS, Smith KS. The role of submesoscale currents in structuring marine ecosystems. Nature Communications. 2018;**9**:4758. DOI: 10.1038/s41467-018-07059-3

[44] Robinson KL, Sponaugle S, Luo JY, Gleiber MR, Cowen RK. Big or small, patchy all: Resolution of marine plankton patch structure at micro-to submesoscales for 36 taxa. Science Advances. 2022;7(47). DOI: 10.1126/sciadv.abk2904

[45] Dave AC, Barton AD, Lozier MS, McKinley GA. What drives seasonal change in oligotrophic area in the subtropical North Atlantic? Journal of Geophysical Research: Oceans. 2015;**120**(6):3958–3969. DOI: 10.1002/2015JC010787

[46] Du C, Liu Z, Kao S-J, Dai M. Diapycnal fluxes of nutrients in an oligotrophic oceanic regime: The South China Sea. Geophysical Research Letters. 2017;**44**(22):510–518. DOI: 10.1002/2017GL074921

[47] Rii YM, Karl DM, Church MJ. Temporal and vertical variability in picophytoplankton primary

productivity in the North Pacific Subtropical Gyre. Marine Ecology Progress Series. 2016;**562**:1–18. DOI: 10.3354/meps11954

[48] Zhou K, Benitez-Nelson CR, Huang J, Xiu P, Sun Z, Dai M. Cyclonic eddies modulate temporal and spatial decoupling of particulate carbon, nitrogen, and biogenic silica export in the North Pacific Subtropical Gyre. Limnology and Oceanography. 2021;**66** (9):3508–3522. DOI: 10.1002/lno.11895

[49] Kemp AES, Villareal TA. High diatom production and export in stratified waters-A potential negative feedback to global warming. Progress in Oceanography. 2013;**119**:4–23. DOI: 10.1016/j.pocean.2013.06.004

[50] Buesseler KO, Bianchi D, Chai F, Cullen JT, Estapa M, Hawco N, John S, McGillicuddy DJ, Morris PJ, Nawaz S, Nishioka J, Pham A, Ramakrishna K, Siegel DA, Smith SR, Steinberg D, Turk-Kubo KA, Twining BS, Webb RM, Wells M, White A, Xiu P, Yoon J-E. Next steps for assessing ocean iron fertilization for marine carbon dioxide removal. Frontiers in Climate. 2024;**6**:1430957. DOI: 10.3389/fclim.2024.1430957

[51] Ecopia Marine. ECOPIA_NBCCMS_MRV_v1_0.pdf. 2024. Available from: https://github.com/JohnAllen01/ECOPIA-MRV/blob/main/ECOPIA_NBCCMS_MRV_v1_0.pdf [Accessed: 2025-November-01]

[52] Friedland KD, Mouw CB, Asch RG, Ferreira ASA, Henson S, Hyde KJW, Morse RE, Thomas AC, Brady DC. Phenology and time series trends of the dominant seasonal phytoplankton bloom across global scales. Global Ecology and Biogeography. 2018;**27**:551–569. DOI: 10.1111/geb.12717

[53] Brown L, Sanders R, Savidge G, Lucas CH. The uptake of silica during the spring bloom in the Northeast Atlantic Ocean. Limnology and Oceanography. 2003;**48**(5):1831–1845. DOI: 10.4319/lo.2003.48.5.1831

[54] Allen JT, Brown L, Sanders R, Moore CM, Mustard A, Fielding S, Lucas M, Rixen M, Savidge G, Henson S, Mayor D. Diatom carbon export enhanced by silicate upwelling in the northeast Atlantic. Nature. 2005;**437** (7059):728–732. DOI: 10.1038/nature03948

[55] Cornec M, Claustre H, Mignot A, Guidi L, Lacour L, Poteau A, D'Ortenzio F, Gentili B, Schmechtig C. Deep chlorophyll maxima in the global ocean: Occurrences, drivers and characteristics. Global Biogeochemical Cycles. 2021;**35**. DOI: 10.1029/2020GB006759

[56] Peperzak L, Timmermans K, Wernand M, Oosterhuis S, Woerd H. A mesocosm tool to optically study phytoplankton dynamics. Limnology and Oceanography: Methods. 2011;**9**:232–244. DOI: 10.4319/lom.2011.9.232

[57] Bouman H, Platt T, Doblin M, Figueiras FG, Gudmundsson K, Gudfinnsson H, Huang B, Hickman A, Hiscock M, Jackson T, Lutz V, Mélin F, Rey F, Pepin P, Segura V, Tilstone G, van Dongen-vogels V, Sathyendranath S. Photosynthesis-irradiance parameters of marine phytoplankton: Synthesis of a global data set. Earth System Science Data. 2018;**10**:251–266. DOI: 10.5194/essd-10-251-2018

[58] Honsberg CB, Bowden SG. Photovoltaics Education Website. 2019. Available from: www.pveducation.org [Accessed: 2025-February-22]

[59] Lee Z, Weidemann A, Kindle J, Arnone R, Carder K, Davis C. Euphotic zone depth: Its derivation and implication to ocean-color remote sensing. Journal of Geophysical Research. 2007;**112**. DOI: 10.1029/2006JC003802

[60] Sverdrup HU. On conditions for the vernal blooming of phytoplankton. Ices Journal of Marine Science. 1953;**18**:287–295. DOI: 10.1093/icesjms/18.3.287

[61] Sathyendranath S, Stuart V, Nair A, Oka K, Nakane T, Bouman H, Forget M, Maass H, Platt T. Carbon-to-chlorophyll ratio and growth rate of phytoplankton in the sea. Marine Ecology Progress Series. 2009;**383**: 73–84. DOI: 10.3354/meps07998

[62] Jakobsen H, Markager S. Carbon-to-chlorophyll ratio for phytoplankton in temperate coastal waters: Seasonal patterns and relationship to nutrients: C: Chl for phytoplankton in temperate coastal waters. Limnology and Oceanography. 2016;**61**:1853–1868. DOI: 10.1002/lno.10338

[63] Siegel DA, Buesseler KO, Behrenfeld MJ, Benitez-Nelson CR, Boss E, Brzezinski MA, Burd A, Carlson CA, D'Asaro EA, Doney SC, Perry MJ, Stanley RHR, Steinberg DK. Prediction of the export and fate of global ocean net primary production: The exports science plan. Frontiers in Marine Science. 2016;**3**(22). DOI: 10.3389/fmars.2016.00022

[64] Yool A, Shepherd JG, Bryden HL, Oschlies A. Low efficiency of nutrient translocation for enhancing oceanic uptake of carbon dioxide. Journal of Geophysical Research. 2009;**114**: C08009. DOI: 10.1029/2008JC004792

[65] Oschlies A, Pahlow M, Yool A, Matear RJ. Climate engineering by artificial ocean upwelling: Channelling the sorcerer's apprentice. Geophysical Research Letters. 2010;**37**:L04701. DOI: 10.1029/2009GL041961

[66] Hoffmann LJ, Peeken I, Lochte K, Assmy P, Veldhuis M. Different reactions of southern ocean phytoplankton size classes to iron fertilization. Limnology and Oceanography. 2006;**51**(3):1217–1229. DOI: 10.4319/lo.2006.51.3.1217

[67] Salter I, Schiebel R, Ziveri P, Movellan A, Lampitt R, Wolff GA. Carbonate counter pump stimulated by natural iron fertilization in the polar frontal zone. Nature Geoscience. 2014;**7** (12):885–889. DOI: 10.1038/ngeo2285

[68] Rohr T, Long M, Kavanaugh MT, Lindsay K, Doney S. Variability in the mechanisms controlling southern ocean phytoplankton bloom phenology in an ocean model and satellite observations. Global Biogeochemical Cycles. 2017;**31**:922–940. DOI: 10.1002/2016gb005615

[69] Allen JT. Marine algae are 'taught' the basics of angular momentum. Ocean Dynamics. 2017;**67**:1429–1442. DOI: 10.1007/s10236-017-1094-8

[70] Freilich MA, Poirier C, Dever M, Alou-Font E, Allen JT, Cabornero A, Sudek L, Choi CJ, Ruiz S, Pascual A, Farrar JT, Johnston TMS, D'Asaro EA, Worden AZ, Mahadevan A. 3D intrusions transport active surface microbial assemblages to the dark ocean. Proceedings of the National Academy of Sciences of the United States of America. 2024;**121**(19): e2319937121. DOI: 10.1073/pnas.2319937121

[71] Chen CTA. On the depth of anthropogenic CO_2. In: penetration in

the Atlantic and Pacific Oceans, Oceanol. Acta, Proceedings International Symposium on Equatorial Vertical Motion; 6-10 May 1985; Paris. 1987. p. 97–102

[72] Sabine CL, Key RM, Feely FA, Greeley D. Inorganic carbon in the Indian Ocean: Distribution and dissolution processes. Global Biogeochemical Cycles. 2002;**16** (4):1067. DOI: 10.1029/2002GB001869

[73] Ritchie H, Rosado P, Roser M CO_2 and Greenhouse Gas Emissions. 2023. Available from: https://ourworldindata.org/co2-and-greenhouse-gas-emissions [Accessed: 2025-April-29]

[74] Quay P, Sonnerup R, Munro D, Sweeney C. Anthropogenic CO_2 accumulation and uptake rates in the Pacific Ocean based on changes in the 13 C/ 12 C of dissolved inorganic carbon. Global Biogeochemical Cycles. 2017;**31**:59–80. DOI: 10.1002/2016GB005460

[75] Gurvan M, Bourdallé-Badie Chanut RJ, Clementi E, Coward A, Ethé C, Iovino D, Lea D, Lévy C, Lovato T, Martin N, Masson S, Mocavero S, Rousset C, Storkey D, Müeller S, Nurser G, Bell M, Samson G, Mathiot P, Mele F, Moulin A. NEMO ocean engine. Scientific Notes of IPSL Climate Modelling Center. 2022;**27**. DOI: 10.5281/zenodo.6334656

[76] Yool A, Popova EE, Anderson TR. MEDUSA-2.0: An intermediate complexity biogeochemical model of the marine carbon cycle for climate change and ocean acidification studies. Geoscientific Model Development. 2013;**6**:1767–1811. DOI: 10.5194/gmd-6-1767-2013

[77] Krishnan M, Samandari H, Woetzel J, Smit S, Pacthod D, Pinner D, Nauclér T, Tai H, Farr A, Wu W, Imperato D The net-zero transition: What it would cost, what it could bring. McKinsey & Company, January 2022. 2022. Available from: https://www.mckinsey.com/capabilities/sustainability/our-insights/the-net-zero-transition-what-it-would-cost-what-it-could-bring [Accessed: 2025-December-01]

[78] World Bank (2025) calculated by Viktoras Kulionis, based on the EXIOBASE v3.8.2 database – Processed by Our World in Data. Consumption-based energy use (trade-adjusted). 2025. Available from: https://ourworldindata.org/grapher/change-energy-gdp-per-capita [Accessed: 2025-April-30]

[79] IEA. Projected Costs of Generating Electricity. Paris: IEA; 2015. License: CC BY 4.0. Available from: https://www.iea.org/reports/projected-costs-of-generating-electricity-2015

[80] IEA. Projected Costs of Generating Electricity. Paris: IEA; 2020. License: CC BY 4.0. Available from: https://www.iea.org/reports/projected-costs-of-generating-electricity-2020

[81] Climate Change Committee. The Seventh Carbon Budget. Advice for the UK Government. 2025. Available from: https://www.theccc.org.uk/publication/the-seventh-carbon-budget/ [Accessed: 2025-April-30]

[82] Office for National Statistics. ONS website, methodology, Measuring UK greenhouse gas emissions. 2024. Available from: https://www.ons.gov.uk/economy/environmentalaccounts/methodologies/measuringukgreenhousegasemissions [Accessed: 2025-April-29]

[83] Department for Environment Food & Rural Affairs. Carbon footprint for the UK and England to 2021 (updated 15 May 2024). 2024. Available from: https://www.gov.uk/government/statistics/uks-carbon-footprint/carbon-footprint-for-the-uk-and-england-to-2019 [Accessed: 2025-April-30]

[84] Department for Energy Security & Net Zero. Traded carbon values used for modelling purposes, 2024. 2024. Availalable from: https://www.gov.uk/government/publications/traded-carbon-values-used-for-modelling-purposes-2024/[Accessed: 2025-April-30]

[85] Trading Economics. EU Carbon Permits. 2025. Available from: https://tradingeconomics.com/commodity/carbon [Accessed: 2025-April-30]

[86] Pareto V. Cours D'économie Politique (In Two Volumes), Vols. **1, 2**. In: Rouge F, Pichon F. Lausanne. Paris Librairies de les Universtes 1896–1897. Available from: https://archive.org/details/fp-0148-1

[87] Box GEP, Meyer RD. An analysis for unreplicated fractional factorials. Technometrics. 1986;**28**(1):11–18. DOI: 10.1080/00401706.1986.10488093

[88] Cannan E. The origin of the law of diminishing returns, 1813-15. The Economic Journal. 1892;**2**(5):53–69. DOI: 10.2307/2955940

[89] Wilson JD, Andrews O, Katavouta A, de Melo Viríssimo F, Death RM, Adloff M, Baker CA, Blackledge B, Goldsworth FW, Kennedy-Asser AT, Liu Q, Sieradzan KR, Vosper E, Ying R. The biological carbon pump in CMIP6 models: 21st century trends and uncertainties. Proceedings of the National Academy of Sciences. 2022;**119** (29). DOI: 10.1073/pnas.2204369119

[90] Siegel DA, DeVries T, Doney SC, Bell T. Assessing the sequestration time scales of some ocean-based carbon dioxide reduction strategies. Environmental Research Letters. 2021;**16**(10):104003. DOI: 10.1088/1748-9326/ac0be0

[91] Baker CA, Martin AP, Yool A, Popova E. Biological carbon pump sequestration efficiency in the north atlantic: A leaky or a long-term sink? Global Biogeochemical Cycles. 2022;**36** (6). DOI: 10.1029/2021GB007286